Microbiology Research Advances

nova
science publishers
www.novapublishers.com

Microbiology Research Advances

Endophytes: Types, Potential Uses and Mechanism(s) of Action
Pragya Tiwari, PhD (Editor)
2022. ISBN: 979-8-88697-045-6 (Hardcover)
2022. ISBN: 979-8-88697-205-4 (eBook)

Catalase and Its Applications
Kaley Rutherford (Editor)
2022. ISBN: 979-8-88697-421-8 (Softcover)
2022. ISBN: 979-8-88697-433-1 (eBook)

Endotoxins and their Importance
Arif Pandit, PhD and R. S. Sethi, PhD (Editors)
2022. ISBN: 978-1-68507-839-3 (Softcover)
2022. ISBN: 978-1-68507-913-0 (eBook)

Listeria monocytogenes: Microbiology, Sites of Infection and Treatment
Raghu Vishweswaraiah, PhD
and Faizan Ahmed, PhD (Editors)
2022. ISBN: 978-1-68507-883-6 (Hardcover)
2022. ISBN: 979-8-88697-087-6 (eBook)

Understanding Antibiofilm Activity
Iram Liaqat, PhD (Editor)
2022. ISBN: 978-1-68507-927-7 (Hardcover)
2022. ISBN: 979-8-88697-005-0 (eBook)

More information about this series can be found at
https://novapublishers.com/product-category/series/microbiology-research-advances/

Qiang-Sheng Wu
Ying-Ning Zou
Yue-Jun He
and Nong Zhou
Editors

New Research on Mycorrhizal Fungus

nova science publishers
www.novapublishers.com

DOI: 10.52305/GLXN2905.

Library of Congress Cataloging-in-Publication Data

ISBN: 979-8-88697-637-3

Published by Nova Science Publishers, Inc. † New York

Contents

Preface

Mycorrhizal fungi are one of the oldest microorganisms that evolved about 400 million years ago from lichenic symbiotic fungi, along with the evolution of plants, and maintained a reciprocal symbiotic relationship with plants. In 1885, the German forester Albert Bemhard Frank observed the colonization of truffles on forest trees and first proposed the concept of mycorrhiza, believing it to be a beneficial association. Subsequently, mycorrhizal research has been developed extensively. Early mycorrhizal studies mainly investigated the colonization characteristics and population diversity in agroforestry in the field, while the middle-term mycorrhizal studies mainly revolved around pot experiments to reveal mycorrhizal fungi in improving plant growth as well as nutrients, especially P acquisition. Nowadays, mycorrhizal symbiotic relationships and their potential functions are revealed by means of molecular techniques. Some new studies on mycorrhizal fungi have also progressed considerably with the help of tools and techniques such as histology and various advanced microscopes.

This book invites mycorrhizal researchers from around the world to write new advances on mycorrhizal fungi and includes nine chapters. Among them, Chapter 1 provides an overview of the germination of arbuscular mycorrhizal (AM) fungal spores from an *in vitro* culture and the effects of secretions and fatty acids as an energy source, and also explored the possibility of the bacteria *Paenibacillus validus* as a biostimulant for AM fungi.

Chapter 2 focuses on exploring the role of biochar in soil and the effects on agriculture when combined with AM fungi, such as the enhancement of glycosides, secondary metabolites and endogenous hormones, nutrient flow and antioxidant enzyme systems.

Chapter 3 focuses on the feedback effects of AM fungi on plant invasion, a global environmental, economic and social problem, and outlined main mechanisms of AM fungi in competition between invasive and native plants, providing a constructive theory for AM fungi to prevent and control harmful invasive plants and maintain ecosystems.

Chapter 4 presents results on the field application of AM fungi in agricultural production in Cuba, along with analyzing potential factors and suggesting the incorporation of mycorrhizal inoculant management into nutrient supply programs and techniques for different crops on various soil types.

Chapter 5 discusses the bioremediation function of AM fungi in soil contamination caused by pollutants in rapid urbanization and industrialization and outlined the role of AM fungi in the sustainable management of soils.

Chapter 6 summarizes the diversity of AM fungi in mangroves from different regions of the world.

Chapter 7 outlines the *in vitro* culture of AM fungi and how pears are mycorrhizalized under *in vitro* culture conditions.

Chapter 8 reviews the role of mycorrhizal fungi in the mineralization of nutrients for plant growth, development and survival.

Chapter 9 outlines the use of AM fungi as an alternative and environmentally friendly method for improving soil health and fertility under the condition of reductions in agricultural production by intensive use of inorganic fertilizers and chemicals, accompanied by the function of AM in improving soil health.

These chapters review new research on mycorrhizal fungi from multiple fields, providing constructive perspectives on mycorrhizal research, and offering readers new research ideas.

Prof. Dr. Qiang-Sheng Wu, Yangtze University, China
Prof. Ying-Ning Zou, Yangtze University, China
Prof. Dr. Yue-Jun He, Guizhou University, China
Prof. Nong Zhou, Chongqing Three Gorges University, China

Acknowledgments

We are sincerely grateful to the contributing authors, who overcame many difficulties to support the editing and publication of this book, and to the editors of Nova Science Publishers for their hard work.

Finally, we would also like to thank the support of the Plan in Scientific and Technological Innovation Team of Outstanding Young Scientists, Hubei Provincial Department of Education (T201604), the National Natural Science Foundation of China (31660156), the Science and Technology Project of Guizhou Province ([2021] General-455), and the Guizhou Hundred-level Innovative Talents Project (Qian-ke-he platform talents [2020] 6004).

Chapter 1

Growth of Arbuscular Mycorrhizal Fungi Independent of a Plant Host

Kasthuri Selvakumar*
Kalaiselvi Thangavel
and Ranjith Sellappan
Department of Agricultural Microbiology,
Tamil Nadu Agricultural University, Coimbatore, India

Abstract

Arbuscular mycorrhizal (AM) fungi are obligatory symbionts that require a host to complete their natural life cycle. It is most likely nature's most significant symbiosis. Many efforts to cultivate *in vitro* have been made recently. The infection and colonization of hosts by arbuscular mycorrhizal fungi rely heavily on root exudates. The hyphal branching of AM fungi is stimulated by exudate portions of host roots. Fatty acids, in addition to being a primary source of root exudate, also function as a stimulant and may aid in the development of AM inoculum *in vitro* production. Specific bacteria also have been shown to increase the interactions between arbuscular mycorrhizal fungi (AMF) and plants, and hence might potentially function as a third partner in this symbiosis. The current studies are to see if AM fungi can grow *in vitro* on media containing diverse sources. It might be a big step forward in the development of AMF that is not dependent on any host plants.

Keywords: arbuscular mycorrhizal fungi (AMF), obligate biotrophy, root exudate, fatty acids, bacteria, asymbiotic condition

---

* Corresponding Author's Email: kasthuri18051999@gmail.com.

In: New Research on Mycorrhizal Fungus
Editors: Qiang-Sheng Wu, Ying-Ning Zou, Yue-Jun He et al.
ISBN: 979-8-88697-637-3

1. Introduction

AM fungi are obligatory symbionts that might need root colonization to complete a life cycle that includes propagules and asexual spores generation. The exterior hyphae of AM fungi act as a stabilizing factor in the creation and development of soil structure. Without the host, AMF have long been suspected to be uncultivable. Though diverse environmental circumstances prompting the onset of germination in different genera and species of AM fungi have been explored, the molecular signals that alleviate spore dormancy and initiate the cell cycle remain unclear. In reality, under the right physical, chemical, and microbiological circumstances, resting spores of many AM fungi species germinate under *in vitro*. Nonetheless, there have been multiple investigations of AM fungal cultivation without hosts. Axenic cultivation of AM fungi has yet to be realized and will need a deeper knowledge of each stage of the fungal life cycle.

2. Diversity of AM Fungi

In the recent two decades, there has been a surge in interest in identifying and comprehending the significance of genetic and functional AMF variety in ecosystem function, as well as the numerous advantages that these fungi may provide to their plant and soil surroundings. Most AMF colonizes a wide range of host plant species, showing that they are not host-specific. AM fungi are thought to be functionally useless because of their poor species diversity and lack of host specificity (Allen et al., 1995). Moreover, AM fungal biomass and community composition vary depending on the habitats, invasive plants, and plant species richness. Small-spored species, such as *Diversipora*, *Dominikia*, *Kamienskia*, *Palaeospora*, and *Paraglomus* spp., or species that sporulate extremely seldom or only seasonally, such as *Acaulospora* spp., make spore identification challenging even for seasoned professionals (Gamper et al., 2009; Silva et al., 2012).

The structure of AM fungal communities was described using ecological indices of diversity such as spore density (SD), species richness (SR), isolation frequency (IF), Shannon-Wiener index (H), and evenness (J) (Simpson, 1949). AMF's functional diversity refers to a wide range of activities linked to plant growth responses to AM fungal invasion. Although the processes that cause functional variety are unknown, the participation of phosphorus (P) and photosynthate exchange between plants and AMF is clear. The phylogeny of AMF has traditionally been based on the morphological properties of spores formed in soil, such as size, color, and cell wall layers. AMF spores, on the other hand, have basic structures with few morphological traits.

Recently, molecular-based technologies in AMF research have allowed direct characterization of AM fungal species in plant roots or soils, revealing that AMF diversity in ecosystems may be larger than previously thought (Husband et al., 2002). The use of the 18S rRNA marker, which is extensively shown in community research on AMF, is also contentious for species-level analysis for some AMF lineages. This was one of the reasons why the AMF communities were studied at the genus level. AMF diversity declines when agricultural land use intensifies, particularly in temperate but also in other climatic zones. Surprisingly, higher AMF diversity was found in agricultural soils as compared to natural environments, particularly when plant growth conditions at the study sites had been greatly enhanced by management methods.

The diversity and richness of AM fungi in agriculture and wild regions are similar, although cropland has a lower relative abundance (Dai et al., 2013). Concerning salinity zones and physiochemical features of soil, the study revealed a greater diversity of arbuscular mycorrhizal fungi in mangrove soil as well as species richness. Arbuscular mycorrhizal fungi were formerly thought to be rare or non-existent in salty soils. However, recent research has shown that AMF may be found in salty and coastal soils. As a result, more study on rhizosphere AMF-related plants is needed.

3. Mass Multiplication of AMF Spores

AMF mass multiplication is highly dependent on root structure and host plant ecosystem. The selection of AM fungal isolates capable of promoting the development of the target plant is also one of the most essential considerations in inoculum generation (Dalpe and Montreal, 2004). Under regulated conditions, the host can impact the variety of mycorrhizal fungi. Monocots and plants with a large root system are excellent hosts for AM fungal proliferation.

Mosse and Hepper were the first to develop an *in vitro* AMF production method (1975). After that, Becard and Fortin (1988) used TDNA-transformed roots of *Daucus carota* to develop the root organ culture system. Several techniques like hydroponics, aeroponics, and soil-less cultivation yield high-quality inoculum with many times the number of propagules as native inoculum. The inoculum has to be in the root zone; seedling roots develop through the inoculum.

The 'funnel method' was used to grow the starter inoculum of each vesicular-arbuscular mycorrhizal (VAM) fungi using a host (Menge and Timmer, 1982). By retaining a funnel on the bottle, a gadget was created. Up to the neck area of the funnel, a tiny amount of sterilized sand and soil mixture was added. The soil mixture's composition is important for optimal mycorrhization and sporulation. The host plant development should be aided by the addition of substrates. The substrates should be chosen in such a way that it has no negative impact on the inoculum. The roots of seedlings were examined after 45 days to check if colonization had occurred (Phillips and Hayman, 1970). For spore quantification, a soil sample was also examined. The entire system, including soil and seedlings, was transferred to larger pots containing sterilized sand: soil for individual spore multiplication utilizing any plant as a host. Before planting VAM spores, they were surface sterilized with 10% sodium hypochlorite or 90% ethanol for 1 min and then completely rinsed with distilled water to remove them.

Root slide analysis was used to assess the total percentage of root colonization (Giovannetti and Mosse, 1980). The ability to penetrate and expand the surface of the host, as well as infiltrate the roots, is an indicator of infectivity. The 'Wet sieving and decanting' procedure was used to extract AMF spores from the soil, which were then counted (Gerdemann and Nicolson, 1963). Spores were extracted and morphotyped based on color, shape, and size using treatments with Polyvinyl Lacto-glycerol (PVLG) and PVLG with Melzer's solution (Schenck and Perez, 1987). The International Culture Collection of Vesicular Arbuscular Mycorrhizal Fungi supplied species descriptions that were used to identify the species (INVAM 2014) (Redecker et al., 2013).

4. Germination of AMF without Plant Host

The majority of AMF species produce spores in the soil, and they can all germinate in the absence of host-derived signals (Hepper and Smith, 1976). Since their spores may germinate and flourish in the absence of host plants supplied with feasible physical, chemical, and microbiological conditions. As a result, host-derived signals are not required for AMF spore germination. In nature, the life cycle of the AM fungi is as follows (Fig. 1). Resting spores are germinated by AM fungi, and germ tubes emerge from the spores and expand into the soil.

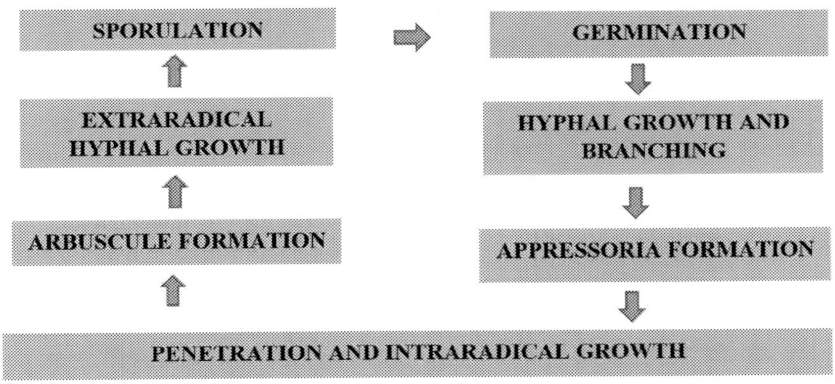

Figure 1. The process of arbuscular mycorrhizal fungal colonization in roots

4.1. Influence of Root Exudates on Germination of AMF Spores

The addition of root exudates has been shown to greatly increase both AM fungal spore germination and hyphal development. According to certain findings, some flavonoids can increase hyphal branching, while others have little impact (Nagahashi and Douds, 2000; Scervino et al., 2005). Flavonoids, like quercetin and quercitrin, have been shown to promote spore germination and/or hyphal development in AM fungal spores (Poulin et al., 1993). The proportion of spore germination was shown to be unaffected by root exudates from nonhost plant species (El-atrach et al., 1989). Mosse's early study demonstrated that AMF may germinate in axenic culture without the presence of a host (Mosse, 1959). Since the publication of this paper, no definitive information on the impact of host roots or root exudates on spore germination has been given.

Exudates from the roots are important in controlling plant–AM fungal interactions and changing the soil microbial population (Hugoni et al., 2018). Amino acids, organic acids, sugars, phenolics, and other secondary metabolites make for a large portion of the variety of root exudates (Rougier, 1981). Exudates of high molecular weight, such as mucilage and proteins, are less diversified, but they frequently make up a significant component of the root exudates. Plants are often cultivated in nutrient solution with subsequent collection of root exudates to reduce mechanical injury to roots (Personeni et al., 2007). Exudates from the roots are often collected at or near room temperature (Ferguson and Menge, 1982). The P nutrition of the host plant has a direct impact on the quality of root exudates and their stimulatory effects on AM fungal development. Elias & Safir (1987), Nair et al., (1991), and Tawaraya et al., (1998), found that exudates generated by plants with adequate P feeding are less beneficial to AM fungi than exudates produced by plants under P stress. When spores are placed on an agar-containing medium, they germinate and generate runner hyphae. After a few weeks, however, the hyphae septate and the cytoplasm retracts, and the growth ends. In vitro studies have demonstrated that volatile root exudates encourage Gigaspora gigantea germ tubes to develop towards the roots of the AM host plant (Koske and Gemma, 1992). These findings raise the question of whether root exudates play a role in VAM infection and if so if they are involved in signaling processes before communication between the two symbionts.

4.2. Fatty Acids as a Carbon and Energy Source in AMF

AM fungi offer minerals to their hosts via hyphal networks in the soil, in exchange for carbon sources obtained from plant photosynthates. Their growth needs a lot of energy and carbon-building components, which are largely delivered to the expanding hyphal tips as lipid droplets and glycogen. AM fungi are obligatory biotrophic fungi that rely on carbohydrates generated from the host, such as sugars and lipids (Keymer et al., 2017; Luginbuehl et al., 2017). Fatty acids have recently been shown to enhance AM fungal growth and sporulation in asymbiotic condition. Infection-competent secondary spore production in *R. irregularis* was promoted by palmitoleic acid asymbiotically (Kameoka et al., 2019). Lipids have recently been discovered to be another plant-derived carbon source. Since there has never been any evidence of fatty acid production in extraradical hyphae, lipids were supposed to be generated in intraradical hyphae and then transported to extraradical hyphae and spores (Bago et al., 1999). Triacylglycerol (TAG) is the most common storage form.

Furthermore, AMF spores created by symbiotically providing palmitoleic acid or myristate were smaller than those generated, and their effectiveness as inoculum is mainly unknown (Sugiura et al., 2020). Outside of mycorrhizal roots, fungal spores store carbon mostly as lipids, with very little quantities of glucose. Although the transfer of nutrients from the fungus to the host has been widely investigated, the transfer of carbon from the host to the fungus, as well as its subsequent metabolism and transport, remains a mystery. Due to the lack of genes involved in de novo fatty acid production in the AM fungal genomes that have been sequenced so far, fatty acids are regarded as critical nutrients for the survival of AM fungi. Exogenous fatty acids are required for the development of AM fungi. Fatty acid desaturases and elongates encoded by genes present within their genomes and synthesized in intraradical hyphae allow AM fungi to use fatty acids once they have been taken up (Wewer et al., 2014).

4.3. The Bacteria, *Paenibacillus validus,* as an AM Fungal Growth Stimulator

AM fungi have long been thought to be uncultivable in the absence of a host. The co-culture of the AMF *Rhizophagus irregularis* and the mycorrhiza-helper bacterium *Paenibacillus validus*, on the other hand, proved that AM fungi may complete their life cycle without the presence of a host (Hildebrandt et al., 2006).

Specific bacteria have been shown to increase AMF-plant interactions and hence might potentially function as a third partner in this. The bacteria-derived compounds are expected to operate as stimulants or nutrition for fungal sporulation, but the molecules responsible have yet to be identified. Diffusible chemicals generated from bacteria are thought to trigger hyphal development. In the AMF, a trisaccharide found in bacterial cultures enhanced hyphal growth but not hyphal branching or sporulation (Hildebrandt et al., 2006).

In the spores and hyphae of the AM fungi *Gigaspora margarita*, a bacterium belonging to the genus *Burkholderia* was discovered. The nitrogenase genes *nifHDK* and *nifH* are present in this bacterium, and the *nifH* gene is expressed in spores (Bianciotto et al., 2003). *Paenibacillus bacteria* have also been found living intracellularly in the ectomycorrhizal fungi *Laccaria bicolor* (Bertaux, 2003). Co-cultivation of AM fungi with a variety of microorganisms may be a means to preserve genetic diversity and function by activating AM fungal genes that would otherwise go unnoticed owing to a lack of environmental stimulation. *Paenibacillus* sp from *Sorghum bicolor* roots enhanced mycorrhizal colonization and inhibited *Phytopthora parasitica* mycelium development *in vitro* (Budi et al., 1999). Mycorrhizal fungal development isn't entirely reliant on plants, but the fact that *P. validus* isolates can stand in for the plant is a step toward better understanding the symbiotic partners' interactions.

5. Future Perspectives and Conclusion

In comparisons of *in vivo* and *in vitro* systems for the formation of spores, the axenic systems would come out on top. Consequently, it will be vital to assess the inoculum potential of various generations of AMF spores in monoxenic environment cultures. To completely comprehend signaling and recognition mechanisms in arbuscular mycorrhiza, more research is needed to pinpoint which exudate molecules are involved in hyphal branching or hyphal tip inhibition.

Under *in vitro* circumstances, the AMF *G. intraradices* produces a sophisticated extraradical hyphal network that colonizes the substrate widely. Surface sterilizing of AMF spores is one of the most significant steps since it is essential for spore germination and the development of the mycelia network. Different specialized structures are formed, and their creation and distribution may be influenced by precise morphogenetic control.

Future research will hopefully provide more light on the development of AM extraradical mycelium, which is one of the most essential but little-known components of arbuscular mycorrhiza.

References

Allen, E. B., Allen, M. F., Helm, D. J., Trappe, J. M., Molina, R., & Rincon, E. (1995). Patterns and regulation of mycorrhizal plant and fungal diversity. *Plant and Soil*, *170*(1), 47-62.

Bago, B., Pfeffer, P. E., Douds Jr, D. D., Brouillette, J., Bécard, G., & Shachar-Hill, Y. (1999). Carbon metabolism in spores of the arbuscular mycorrhizal fungus *Glomus intraradices* as revealed by nuclear magnetic resonance spectroscopy. *Plant Physiology*, *121*(1), 263-272.

Bécard, G., & Fortin, J. A. (1988). Early events of vesicular–arbuscular mycorrhiza formation on Ri T-DNA transformed roots. *New Phytologist*, *108*(2), 211-218.

Bertaux, J., Schmid, M., Prevost-Boure, N. C., Churin, J. L., Hartmann, A., Garbaye, J., & Frey-Klett, P. (2003). In situ identification of intracellular bacteria related to Paenibacillus spp. in the mycelium of the ectomycorrhizal fungus *Laccaria bicolor* S238N. *Applied and Environmental Microbiology*, *69*(7), 4243-4248.

Bianciotto, V., Lumini, E., Bonfante, P., & Vandamme, P. (2003). 'Candidatus *Glomeribacter gigasporarum*' gen. nov. sp. nov., an endosymbiont of arbuscular mycorrhizal fungi. *International Journal of Systematic and Evolutionary Microbiology*, *53*(1), 121-124.

Budi, S. V., van Tuinen, D., Martinotti, G., & Gianinazzi, S. (1999). Isolation from the *Sorghum bicolor* mycorrhizosphere of a bacterium compatible with arbuscular mycorrhiza development and antagonistic towards soilborne fungal pathogens. *Applied and Environmental Microbiology*, *65*(11), 5148-5150.

da Silva, D. K. A., Pereira, C. M. R., de Souza, R. G., da Silva, G. A., Oehl, F., & Maia, L. C. (2012). Diversity of arbuscular mycorrhizal fungi in resting and dunes areas in Brazilian Northeast. *Biodiversity and Conservation*, *21*(9), 2361-2373.

Dai, M., Bainard, L. D., Hamel, C., Gan, Y., & Lynch, D. (2013). Impact of land use on arbuscular mycorrhizal fungal communities in rural Canada. *Applied and Environmental Microbiology*, *79*(21), 6719-6729.

Dalpe, Y., & Monreal, M. (2004). Arbuscular mycorrhiza inoculum to support sustainable cropping systems. *Crop Management*, *3*(1), 1-11.

El-Atrach, F., Vierheilig, H., & Ocampo, J. A. (1989). Influence of non-host plants on vesicular-arbuscular mycorrhizal infection of host plants and spore germination. *Soil Biology and Biochemistry*, *21*(1), 161-163.

Elias, K. S., & Safir, G. R. (1987). Hyphal elongation of *Glomus fasciculatus* in response to root exudates. *Applied and Environmental Microbiology*, *53*(8), 1928-1933.

Ferguson, J. J., & Menge, J. A. (1982). The influence of light intensity and artificially extended photoperiod upon infection and sporulation of *Glomus fasciculatus* on Sudan grass and root exudation of Sudan grass. *New Phytologist*, *92*(2), 183-191.

Gamper, H. A., Walker, C., & Schüßler, A. (2009). *Diversispora celata* sp. nov: molecular ecology and phylotaxonomy of an inconspicuous arbuscular mycorrhizal fungus. *New Phytologist*, *182*(2), 495-506.

Gerdemann, J. W., & Nicolson, T. H. (1963). Spores of mycorrhizal Endogone species extracted from soil by wet sieving and decanting. *Transactions of the British Mycological Society*, *46*(2), 235-244.

Giovannetti, M., & Mosse, B. (1980). An evaluation of techniques for measuring vesicular arbuscular mycorrhizal infection in roots. *New Phytologist*, 489-500.

Hepper, C. M., & Smith, G. A. (1976). Observations on the germination of Endogone spores. *Transactions of the British Mycological Society*, *66*(2), 189-194.

Hildebrandt, U., Ouziad, F., Marner, F. J., & Bothe, H. (2006). The bacterium *Paenibacillus validus* stimulates the growth of the arbuscular mycorrhizal fungus *Glomus intraradices* up to the formation of fertile spores. *FEMS Microbiology Letters*, *254*(2), 258-267.

Hugoni, M., Luis, P., Guyonnet, J., & Haichar, F. E. Z. (2018). Plant host habitat and root exudates shape fungal diversity. *Mycorrhiza*, *28*(5), 451-463.

Husband, R., Herre, E. A., Turner, S. L., Gallery, R., & Young, J. P. W. (2002). Molecular diversity of arbuscular mycorrhizal fungi and patterns of host association over time and space in a tropical forest. *Molecular Ecology*, *11*(12), 2669-2678.

Kameoka, H., Tsutsui, I., Saito, K., Kikuchi, Y., Handa, Y., Ezawa, T., ... & Akiyama, K. (2019). Stimulation of asymbiotic sporulation in arbuscular mycorrhizal fungi by fatty acids. *Nature Microbiology*, *4*(10), 1654-1660.

Keymer, A., Pimprikar, P., Wewer, V., Huber, C., Brands, M., Bucerius, S. L., ... & Gutjahr, C. (2017). Lipid transfer from plants to arbuscular mycorrhiza fungi. *elife*, *6*, e29107.

Koske, R. E., Gemma, J. N., & Flynn, T. (1992). Mycorrhizae in Hawaiian angiosperms: a survey with implications for the origin of the native flora. *American Journal of Botany*, *79*(8), 853-862.

Luginbuehl, L. H., Menard, G. N., Kurup, S., Van Erp, H., Radhakrishnan, G. V., Breakspear, A., ... & Eastmond, P. J. (2017). Fatty acids in arbuscular mycorrhizal fungi are synthesized by the host plant. *Science, 356*(6343), 1175-1178.

Mosse, B., & Hepper, C. (1975). Vesicular-arbuscular mycorrhizal infections in root organ cultures. *Physiological Plant Pathology, 5*(3), 215-223.

Nagahashi, G., & Douds Jr, D. D. (2011). The effects of hydroxy fatty acids on the hyphal branching of germinated spores of AM fungi. *Fungal Biology, 115*(4-5), 351-358.

Nair, M. G., Safir, G. R., & Siqueira, J. O. (1991). Isolation and identification of vesicular-arbuscular mycorrhiza-stimulatory compounds from clover (*Trifolium repens*) roots. *Applied and Environmental Microbiology, 57*(2), 434-439.

Pérez, Y., & Schenck, N. C. (1990). A unique code for each species of VA mycorrhizal fungi. *Mycologia, 82*(2), 256-260.

Personeni, E., Nguyen, C., Marchal, P., & Pagès, L. (2007). Experimental evaluation of an efflux–influx model of C exudation by individual apical root segments. *Journal of Experimental Botany, 58*(8), 2091-2099.

Phillips, J. M., & Hayman, D. S. (1970). Improved procedures for clearing roots and staining parasitic and vesicular-arbuscular mycorrhizal fungi for rapid assessment of infection. *Transactions of the British Mycological Society, 55*(1), 158-IN18.

Redecker, D., Schüßler, A., Stockinger, H., Stürmer, S. L., Morton, J. B., & Walker, C. (2013). An evidence-based consensus for the classification of arbuscular mycorrhizal fungi (Glomeromycota). *Mycorrhiza, 23*(7), 515-531.

Rougier, M. (1981). Secretory activity of the root cap. In *Plant carbohydrates II* (pp. 542-574). Springer, Berlin, Heidelberg.

Scervino, J. M., Ponce, M. A., Erra-Bassells, R., Vierheilig, H., Ocampo, J. A., & Godeas, A. (2005). Flavonoids exhibit fungal species and genus-specific effects on the presymbiotic growth of *Gigaspora* and *Glomus*. *Mycological Research, 109*(7), 789-794.

Simpson, E. H. (1949). Measurement of diversity. *Nature, 163*(4148), 688-688.

Sugiura, Y., Akiyama, R., Tanaka, S., Yano, K., Kameoka, H., Marui, S., ... & Saito, K. (2020). Myristate can be used as a carbon and energy source for the asymbiotic growth of arbuscular mycorrhizal fungi. *Proceedings of the National Academy of Sciences, 117*(41), 25779-25788.

Tawaraya, K., Hashimoto, K., & Wagatsuma, T. (1998). Effect of root exudate fractions from P-deficient and P-sufficient onion plants on root colonization by the arbuscular mycorrhizal fungus *Gigaspora margarita*. *Mycorrhiza, 8*(2), 67-70.

Wewer, V., Brands, M., & Dörmann, P. (2014). Fatty acid synthesis and lipid metabolism in the obligate biotrophic fungus *Rhizophagus irregularis* during mycorrhization of *Lotus japonicus*. *The Plant Journal, 79*(3), 398-412.

Chapter 2

The Combined Use of Arbuscular Mycorrhizal Fungi and Biochar for the Improvement of Soil Biochemistry

Abeer Hashem[1,2]
Qiang-Sheng Wu[3]
Khalid F. Almutairi[4]
Saud S. Aloud[5]
and Elsayed Fathi Abd_Allah[4,*]
[1] Botany and Microbiology Department,
College of Science, King Saud University, Riyadh, Saudi Arabia
[2] Mycology and Plant Disease Survey Department,
Plant Pathology Research Institute, ARC, Giza, Egypt
[3] College of Horticulture and Gardening,
Yangtze University, Jingzhou, China
[4] Plant Production Department,
College of Food and Agricultural Sciences,
King Saud University, Riyadh, Saudi Arabia
[5] Soil Sciences Department,
College of Food and Agricultural Sciences,
King Saud University, Riyadh, Saudi Arabia

Abstract

Biochar alters soil ecosystems and has a direct impact on the symbiotic connection between plants and mycorrhizal fungi. Biochar has been

---

[*] Corresponding Author's Email: eabdallah@ksu.edu.sa.

In: New Research on Mycorrhizal Fungus
Editors: Qiang-Sheng Wu, Ying-Ning Zou, Yue-Jun He et al.
ISBN: 979-8-88697-637-3

suggested to be a multipurpose soil supplement that can boost soil fertility and crop productivity, while reducing nutrient losses and sequestering carbon. The application of biochar to the soil has been reported to significantly improve plant appearance, health, and root morphology. Biochar applications enhance soil enzymatic activity which improves Na and K homeostasis, as well as soil tilth, soil carbon, soil nutrition, microbial activity, and soil water capacity.. The combined application of biochar and arbuscular mycorrhizal fungi (AMF) enhances the level of siderophores, secondary metabolites and endogenous hormones, as well as increasing nutrient transport and antioxidant enzyme capacity, while decreasing chlorophyllase activity. The combined application of AMF and biochar improves root morphology, plant metabolism, and yield.

Soil microbial biomass and the colonization rate of AMF are also enhanced by the application of biochar. These contrasting, strong effects suggest that plant response to biochar application is largely influenced by the soil matrix, as well as by microbes such as AMF. The interaction of biochar with arbuscular AMF, which are significant root symbionts of many terrestrial plants, including most crops, has yet to be investigated. Biochar-induced inhibition of plant growth and mineral nutrition is also considerably alleviated through its interaction with AMF, primarily by enhancing plant phosphorus uptake.

Based on the reported effects of biochar, the present chapter reports on the effect of both AMF and biochar on soil biochemistry in different soil types, as well as the combined effect of these agents on root morphology, organic matter (OM) content, and several plant characteristics. The chapter also highlights the more dynamic studies on the old and new uses of biochar and its synergistic interaction with AMF which alters the physicochemical properties of soils.

Keywords: biochar, AMF, soil, biochemistry, plants growth

1. Introduction

Biochar is a by-product of carbon (C) rich components created when organic wastes are pyrolyzed (Söderberg & Eckerberg, 2013) under a low oxygen atmosphere (Lehmann et al., 2011). Carbon sequestration, mitigation of soil contamination, improved soil nutrient capacity, enhanced crop productivity, and environmental restoration are all considered emerging uses of biochar (Maraseni, 2010; Spokas et al., 2012). Biochar has a strong ability to alter the physio-chemical properties of many different soil types (Kuppusamy et al.,

2016), as well improve soil carbon levels, enhance microbial activity, improve nutrient retention (Sarthou et al., 2014), reduce greenhouse gas (GHG) emissions (Chang et al., 2016), and increase the sorption of pollutants (Cheng et al., 2017; Kuppusamy et al., 2016; Williams et al., 2015). Biochar applications have been recommended to boost soil fertility, water absorption, and carbon sequestration (CS), as well as to counteract anthropogenic CO2 emissions (Roberts et al., 2010). These ideas are based on the presence of terra preta soils in the Amazon or other parts of the tropics where biochar and organic fertilizers have been used to improve crop nutrition and increase yields of crops grown on largely unfertile, heavily weathered, and predominantly acid soils (Mao et al., 2012).

Biochar applications have been suggested to improve soils by enhancing macro-nutrient accessibility [particularly nitrogen (N) and phosphorus (P)], limiting water loss by evaporation, mitigating pH extremes in tropical soils, mitigating heavy metal toxicity, such as Al+3 and Mn+2, and balancing the cation exchange capacity (CEC) of soils (Alling et al., 2014; Gul & Whalen, 2016). Biochar amendments can improve water holding capacity in sandy soils and aeration in heavy clays by altering soil porosity (Johnson et al., 2016; B. S. Mickan et al., 2016). Unlike in tropical soils, where biochar applications can produce genuinely dramatic results, biochar applications in temperate soils have often resulted in far lesser benefits on plants and soil quality (Jeffery et al., 2017; Koide, 2017; Labarrere et al., 2011). This has been attributed in part to the fact that the majority of studies in temperate soils have used fresh biochar, however, the biggest advantages in the use of biochar has been suggested to come from the use of weathered biochar weathering and its interaction with minerals in the soil over long periods of time (Atkinson et al., 2010; Ren et al., 2018). Furthermore, various forms of biochar have been reported to effectively mitigate or improve different soil qualities, such as pH extremes, textural restrictions, and the bioavailabity of toxic metals.

The impact of the biochar is affected by various parameters, such as temperature and oxygen availability, in its preparation (Butnan et al., 2015; Kloss et al., 2012). As a result, the type of biochar used, as well as its application rate, has a significant impact on the ability of biochar to affect various soil parameters, such as microbes, organic matter, and nutrients, which directly affect the ability of a plant to tolerate and thrive under biotic and abiotic stress conditions (Luo et al., 2017; Shen et al., 2016; Steinbeiss et al., 2009). A recent study conducted in China suggested that the application of large amounts of biochar (30-50 tha-1) along with AMF enhances the level of saprotrophic fungal substances, as well as bacterial activity, which improves the physical and chemical properties of soils (Luo et al., 2017).

A meta-analysis of studies conducted with biochar indicated that biochar significantly increases the abundance of gram (+) and (-) bacterial, bacterial biomass, and total biomass in treated soils (Zhang et al., 2018). Song et al. (2014) suggested that the dynamics of ammonia oxidizers in the soil were altered by biochar application as a result of its impact on ammonia-oxidizing bacteria. Community composition and the abundance ammonia-oxidizing bacteria in the soil were increased in soils treated with biochar. Different mechanisms are associated with the enhancement of soil properties, which involve the availability of labile C and N, which are also affected by biochar applications and soil pH changes. PAHs and flavonoid compounds can also modulate the composition and abundance of soil microbes (Zheng et al., 2018).

Our current understanding of the effects of biochar and AMF on soil improvement suggests that a combination of biochar and AMF improves the breakdown, adsorption, and leaching of toxic compounds from soils. The present review provides an overview of how AMF inoculation of soils in combination with the application of biochar application enhances the physical and chemical properties of soils, as well as increases soil organic matter content and microbial activity. Additionally, we also provide qualified predictions on the combined effects of biochar and AMF applications on soil processes and the health, appearance, and yield of field-grown crops.

2. Mechanism of Actions

The combined application of biochar and AMF alters the chemical and biochemical environment of soil, which allow plants to more readily absorb nutrients, as well as absorb CO_2 from the atmosphere for photosynthetic activity (Li and Cai, 2021). Biochar has also been reported to have the ability to protect beneficial soil microbes (especially bacteria) from predators by creating suitable refugia (Lehmann et al., 2011). The mycelia of AMF inhibit the movement of heavy metals and release glomalin, which alters the pH of the soil (Spagnoletti et al., 2017). Soil pollutants also affect AMF activity, however, and influence nutrient availability in soils (Treseder & Allen, 2002).

More than half of all current plant species, including many key crop and grassland species, have a symbiotic interaction with AMF. This symbiosis is responsible for up to 100% of the phosphorus (P) uptake in plants, as well as a large portion of the nitrogen (N) uptake by plants, especially from organic N sources (Hodge & Fitter, 2010; Hodge & Storer, 2015; Smith & Smith, 2012; Smith et al., 2004). AMF hyphae have been shown to penetrate biochar particles and acquire P that has been transiently adsorbed onto biochar surfaces (Hammer et al., 2015). Furthermore, the addition of biochar addition to water deficient soils boosts the formation of AMF extraradical hyphae, relative to unamended soil (B. S. Mickan et al., 2016). Changes in microbial abundance and community structure may have an impact on biochar mineralization, as well as other soil C mineralization (Wardle et al., 2008). Different types of labile compounds, such as volatile organic compounds (VOCS), may boost microbial activity in soils after biochar has been applied (Steiner et al., 2008).

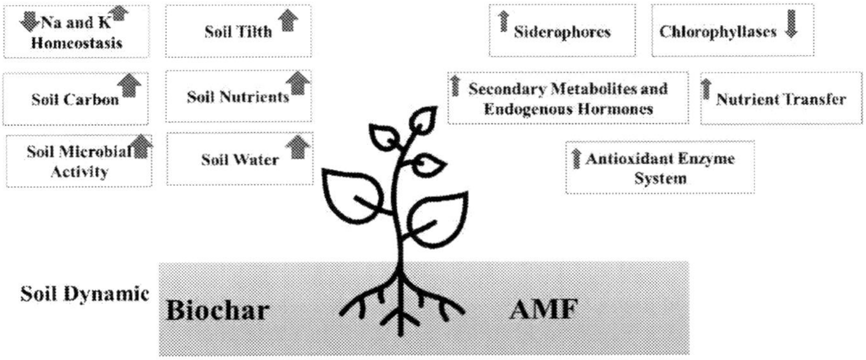

Figure 1. This schematic diagram represents combined effects on soil properties as well as plant growth using certain adoptive mechanisms.

Despite the increased mineral fertility of such soils, compared to unamended soils, an analysis of microbial communities in African Dark Earths revealed no significant influence of biochar on indigenous AMF (despite a considerably larger fungal to bacterial ratio) (Camenzind et al., 2018). The lack of a significant effect of biochar on mycorrhizal abundance is consistent with an earlier meta-analysis of the effects of biochar (Biederman & Harpole, 2013).

Biochar, however, may still have an indirect impact on the AMF community and its function by changing abiotic soil properties such as pH (Hazard et al., 2013; Jansa et al., 2014), modulating the composition and activity of the overall soil microbial community (Luo et al., 2017), and/or affecting biological soil processes, such as nitrification (Gul & Whalen, 2016;

Luo et al., 2017; Schulzke, 2019). These processes could have a significant impact on AMF functioning and warrant further investigation and discussion (Bornø et al., 2018; Bukovská et al., 2018; Cheng et al., 2012). In fact, the potential significance of AMF in OM breakdown was noted in the cited studies. The combined synergistic effect of biochar and AMF on the biological and physicochemical properties of soils and on plant growth is presented in Figure 1.

2.1. The Effect of Biochar Applications on the Soil Microbiome

Biochar is a good sorbent because of its high specific area and microporous structure (Khorram et al., 2015; Srinivasan & Sarmah, 2015), properties that could be used to limit pesticide residues (Jones et al., 2011; Yu et al., 2009) and pesticide leaching into groundwater (Marín-Benito et al., 2013). The use of biochar as a targeted method for managing soil biota is gaining interest, although unintentional changes in soil biota due to biochar use are also a source of concern. This is an important topic of research because the health and diversity of the soil microbial community plays an important role in soil and ecosystem functioning, affecting soil structure and stability, nutrient cycling, aeration, water use efficiency, disease resistance, and C storage capacity (Brussaard, 1997).

Biochar, which has an alkaline pH, can increase the pH of acidic soils, and due to its vast surface areas and micropores, biochar can significantly adsorb soil pollutants and release enzymes and chemicals that mitigate soil pollution (Ahmad et al., 2014). Crop straw, livestock and poultry manure, and other biomass waste are pyrolyzed at high temperatures under anoxic conditions to produce biochar, resulting in a carbon-rich, stable solid product. Biochar also has a significant impact on phosphorus recycling (Efthymiou et al., 2018; Zwetsloot et al., 2016). Various studies suggest that biochar can be used as a soil amendment for managing pathogens, adsorbing metals and other toxins, and as microbial inoculant carriers (Lehmann et al., 2011). It is evident that biochar can improve the sorption properties, pH, and physical characteristics (pore shape, surface area, and mineral content) of soils, as well as the activity of soil biota.

It is feasible to suggest, based on observations of microbial dynamics, that resource utilization could be improved by biochar because of the co-location of various resources in and around biochar. Increased soil biota abundance is presumably the result of absorption of nutrients and the inactivation of growth-inhibiting chemicals. The release of a range of organic compounds from new biochar, however, may be responsible for increases or declines in soil biota abundance and activity. The reasons for variations in microbial abundance may vary for different microbes. Warnock et al. (2010) reported that the two most frequent forms of mycorrhizal fungi, AMF and ectomycorrhizal (EMF) are often positively influenced by the application of biochar.

Nitrogen (N) is an inorganic compound required by plants to function and grow (Nishida et al., 2018). The application of nitrogen in the form of NH^{4+} induces soil nitrification and the indirect release of H^+, potentially increasing the rate of soil acidification (Zhao et al., 2020). Due to the dissolution of the organophosphate chemical link in woody tissues during pyrolysis, biochar retains a substantial quantity of soluble phosphate (Gundale & DeLuca, 2006). Therefore, adding biochar to the soil can boost the amount of soluble phosphate available in soils. Extracellular enzymes mediate organophosphate hydrolysis, with the most important factors regulating phosphorus mineralization being enzyme activity and microbial biomass (Gul & Whalen, 2016).

The microporous structure of biochar and readily oxidizable organic carbon content create a favorable environment for microorganism growth. One study indicated that application of 3.5 t/ha biochar for 20 days increased the microbial biomass carbon (MBC) content by threefold, while phosphatase activity increased by 19%, and phosphorus mineralization was enhanced. Vassilev et al. (2013) reported that the nutrients in biochar coal ash induced microbes to secrete more dissolved organo-phosphates. Biochar was also found to affect phosphorus availability by influencing root growth. Biochar application to soils boosted root biomass in barley by 29% and the contact area between the rhizosphere and the soil by 14%, resulting in a 27% increase in root phosphorus uptake (Prendergast-Miller et al., 2014).

Biochar and AMF both improve plant growth but in different ways. The physical and chemical properties of soils are improved by biochar. For example, nutrient retention and availability of soils are improved by the ability of biochar to increase both water holding capacity and CEC. When biochar is added to soil crop growth and stress tolerance are improved (Li and Cai, 2021). Biochar greatly increased the colonization rate of AMF and synergistically increased phosphorus absorption (de Figueiredo et al., 2019). Notably, the phosphorus concentration of biochar was not altered by pyrolysis temperature (de Figueiredo et al., 2019).

As previously indicated, organophosphate linkages in biomass woody tissue are broken during pyrolysis, releasing soluble forms of phosphorus that persist in the biochar. The application of biochar increases the level of carbon and soluble (P) in the soil, allowing plants to absorb a greater amount of P (D. Zhang et al., 2016). The addition of biochar was reported to significantly reduce the simazine breakdown and simazine content in soil leachate, relative to unamended soil. Biochar application also lowered ^{14}C activity in plants while increasing ^{14}C activity in the soil (Cheng et al., 2020). Biochar treatment improves the adsorption capacity of soils by increasing soil pH, carbon priming, its microporous structure, and by regulating soil microbial community biomass and activity (Söderberg & Eckerberg, 2013).

2.2. The Impact of AMF on Soil Quality and Plant Growth

AMF are soil microorganisms that create a mutual association with the roots of most higher plant species (Nelly et al., 2019). AMF accumulates metal ions and enhance soil fertility by increasing the organic matter content of soils (S. Andrade et al., 2010). The establishment of an AM association plays a critical role in plant development and nutrition (Bukovská et al., 2018; Gryndler et al., 2018), but does not affect the acidity of soils (Paymaneh et al., 2018). Paymaneh et al., (2018) have also reported that both non-amended acidic soils and the amendment of acidic soils with biochar can suppress plant growth and reduce plant nutrient availability. The addition of AMF, however, completed eliminated the adverse effects of the addition of biochar in acid soils.

The addition of biochar to alkaline soils had no negative impact on plant growth with or without the addition of AMF (Paymaneh et al., 2018). These results are in agreement with the premise that biochar has neutral to favorable impacts on plant growth and/or mineral nutrition under different soil conditions (Brodowski et al., 2006). The potential of AMF hyphae to boost water and nutrient uptake by plants due to the symbiotic interaction between plants and AMF (Smith et al., 2010) is a plausible mechanism that may explain the variable impact of biochar on plant growth. The addition of a biochar compost mixture to the soil boosted average crop productivity by up to 10%, (Jeffery et al., 2011), and increased total organic carbon and nitrogen content. Results reported in previous studies have documented the transfer of available N from AMF mycelia to plants (Bukovská et al., 2018). AMFs are assumed to be fully reliant on other bacteria to liberate mineral nitrogen from organic nitrogen sources. Other studies have found that AMF can limit microbial interference in the soil by co-opting N and P in the soil rhizosphere, or via allelopathy (direct or indirect) (Bukovská et al., 2018; Koide, 2017).

AMF hyphae also enhance the surface area accessible for phosphorus absorption and also emit exudates that mineralize and solubilize phosphorus (Koide & Kabir, 2000). The low pyrolysis temperature used to create biochar is unlikely to volatilize the P in the plant tissues utilized to make biochar since phosphorus cannot volatilize at temperatures below 500–700 C (Adhikari et al., 2019). The mycelia of AM fungi recognize and translocate Cd towards roots, and the immobilization of AMF on biochar particles can reduce the exposure and transport of contaminants to plant roots (Zhang et al., 2015). The uptake of phosphorus by plant roots is aided by soil microorganisms that promote root development and phosphorus distribution between inorganic and organic pools (Richardson, 2001).

AMF are found on the surface of roots, as well as in cortical and surrounding epidermal tissues of roots and have a symbiotic relationship with most terrestrial plants. AMF hyphae are thinner than root hairs, allowing them to reach soil micropores that plant root hairs cannot access, where they can obtain essential nutrients (mostly nitrates and phosphates) needed for plant growth, which they subsequently deliver to their host plant in exchange for carbohydrates. As a result, mycorrhizal symbiosis has a major impact on plant development and adaptability (van der Heijden et al., 2015). Even though AMF can considerably increase phosphorus availability, their ability to produce phosphatase and activate organophosphates is limited (Tisserant et al., 2013).

Therefore, AMF optimize phosphorus use in other ways. First, AMF can quickly release carbohydrates produced by photosynthesis to the rhizosphere, stimulating microbial growth and abundance, resulting in the enhanced secretion of phosphatase and the activation of organophosphates, which indirectly increases phosphorus absorption by plant roots (L. Zhang et al., 2016). Second, AMF can alter the morphology of plant roots in a manner that facilitates their ability to absorb phosphorus (Cosme & Wurst, 2013). For example, Zhang et al., (2016), reported that AMF can increase the absorption area of plant roots by constructing a hyphal network that can absorb phosphorus from soil pores and transmit it to the roots of host plants.

2.3. The Collective Benefit of AMF and Biochar Application to Soils

The addition of biochar combined with AMF have recently received increased interest as a solution for reducing heavy metal contamination in soils (Hu et al., 2013; Vejvodová et al., 2020). This is due to the porous structure, wide surface area, and abundance of micropores in biochar (Ahmad et al., 2014; Cao & Harris, 2010). Several studies have indicated that the addition of AMF improves the physical and chemical properties of soils and also binds heavy metals in their mycelia (L. R. Andrade et al., 2010; Göhre & Paszkowski, 2006). Glomalin/fungal polyphosphates and metallothioneins also can chelate metals (Andrade et al., 2010; Göhre & Paszkowski, 2006). Despite the recognition of the various benefits of AMF and biochar, few studies have investigated the interaction and possibly synergistic effects of the two components working together (M. Mickan et al., 2016). The collective impact of the combined use of AMF and biochar could be used as a method to limit the presence of pollutants in soils, which would ultimately affect plant yield. Application of hydrocarbon increases the microbial activity-dependent in soil pH and also help to the undiscovered microbial based reduction of substrate or electron acceptors (Rillig et al., 2010).

A few studies have reported on the reclamation of soils by the combined use of AMF and biochar (Cheng et al., 2020). One study reported on the removal of heavy metals from the soil using a combined AMF and biochar approach (Liu et al., 2018). The results indicated that Cd immobilization could be increased and Cd phyto-availability reduced by the combined use of AMF and biochar, resulting in enhanced plant growth and lower Cd toxicity in heavy metal contaminated soil (Liu et al., 2018). Interactions between biochar and N fertilizer might improve N retention in soils, resulting in higher nutrient efficiency and crop yields (Ibrahim et al., 2020). Combining BC and N fertilizer could also potentially improve the stability of the microbial community, which would have a favorable impact on the biological activity in soils (Song et al., 2020). The high buffering capacity of biochar and its ability to gradually increase the pH of soils could help to prevent nitrification, which in turn would lower the amount of accessible Cd in acidic soils (Zhao et al., 2020).

Biochar greatly improved AMF colonization and boosted P levels in shoots and roots, compared to non-amended soils. Another study reported that biochar can be used to release P to wheat (*Triticum aestivum* L.) plants and enhance AMF colonization (Solaiman et al., 2010). P uptake has been demonstrated to be associated with AMF colonization (Blackwell et al., 2015). Biochar appears to enhance mycorrhizal colonization by boosting P levels in chickory and the effect was not impacted by the presence of nitrogen fertilizer (Zhao et al., 2021). N fertilizer reduces AMF colonization in maize and soybeans (*Glycine max* L.), regardless of additional treatments, (Zhang et al., 2020a). These disparities could be due to differences in AMF species, soil fertility, fertilization levels, and sampling methodologies (Tian et al., 2013). AMF inoculation boosted shoot and root biomass, while the addition of biochar increased shoot biomass (Zhao et al., 2021).

AMF promotes nutrient absorption and boosts plant yields in chickpea (*Cicer arietinum* L.) (Farzaneh et al., 2011). Notably, increasing plant yield is the predominant benefit of the application of biochar to soils (Ahmad et al., 2014). Biochar promotes soil fertility in acidic soils by increasing nutrient availability, stabilizing moisture content, enhancing N retention, which improves plant efficiency resulting in improves plant yields (Liu et al., 2013). The liming effect of biochar amendments stabilizes soil pH, which is also beneficial to plant growth (Ahmad et al., 2014). The addition of biochar to the soil, along with AMF inoculation, substantially raised the soil pH Thus, biochar enhances plant growth by increasing soil fertility and pH. In a N fertilization experiment, the biggest increase in shoot biomass was observed in plants growing in a soil treated with AMF+BC+N (Zhao et al., 2021).

Biochar can increase AMF abundance by changing the physicochemical properties of soils, influencing the activity of other microbes, and changing the AMF signaling dynamics that protect AMF (Warnock et al., 2007). The abundance of functional groups of chemical compounds can be modified by N fertilizer treatment, while AMF inoculation can improve plant production and biomass stability (Kang et al., 2020). Combining biochar with N fertilizer also enhances plant development and biomass production (Tian et al., 2016). Notably, AMF are able to absorb some of the inorganic nitrogen and retain it in fungal mycelia in the soil (Tomè et al., 2015), resulting in a reduction in the transport of inorganic nitrogen from the soil to plant roots, thus, lowering root biomass.

AMF are important for plant nutrient uptake, particularly for P uptake (Farzaneh et al., 2011). Extracellular phosphatases and phosphate are secreted once and AMF symbiosis is established, and organic acids solubilize P, resulting in enhanced absorption of P by plant roots (Gul & Whalen, 2016). The application of biochar enhances organic P hydrolysis (Gul and Whalen, 2016), which may explain why P levels increase following the application of biochar. Even in the presence of soil micro-communities, AMF symbiosis improves nutrient intake even in the presence of the entire soil microbial community (Farzaneh et al., 2011). The application of biochar was reported to enhance the growth of *Phragmites karka* by improving nutrient uptake (Abideen et al., 2020). Hammer et al. (2014) utilized ^{33}P to confirm that AMF hyphae can infiltrate the micropores of biochar and collect phosphorus. Adding biochar to soils has been demonstrated in several studies to improve soil quality and the rate of AMF colonization (B. S. Mickan et al., 2016; Solaiman et al., 2010). The mechanism underlying this ability could include changes in soil characteristics and increased microbial activity (Warnock et al., 2007). Interactions between the AMF hyphal network and biochar are beneficial to plants because they increase access to nutrient-rich soil locations that would otherwise be inaccessible (Zwetsloot et al., 2016).

3. The Effect of Biochar and AMF On Spore Formation and Enzyme Activity in Soils

Biochar is a carbon-rich substance produced through the pyrolysis of variety of different types of biomasses. It can play an important role in reducing global warming and the atmospheric concentration of CO_2, as well as improve the quality of agricultural soils (Lehmann et al., 2006; Zamin et al., 2020). The application of biochar has been shown to boost microbial activity in soils and improve their physical and chemical properties (El Nahhas et al., 2021; Yamato et al., 2006). Biochar treatments also increase organic matter content in e soils (Ścisłowska et al., 2015) and considerably boost the water-holding capacity, cation exchange capacity, and specific surface area of soils (Laird et al., 2010). Rice husk biochar enhanced organic matter in the soil by 52.94%, relative to a control (Qayyum et al., 2020).

According to several studies, the addition of biochar alone increased the activity of dehydrogenase, alkaline phosphatase, acid phosphatase, acid phosphor-monoesterase, alkaline phosphor-monoesterase, protease, chymo-trypsin, trypsin, phosphohydrolase, lipase-esterase, and esterase enzymes in the soil (Anderson et al., 2011; Trupiano et al., 2017). Biochar also enhances nutrient absorption and availability in soils and has been reported to increase nitrogen, potassium, calcium, magnesium, sodium, and total carbon levels in soils (Saxena et al., 2013; Wang et al., 2014). Jabborova et al., (2020 also reported that biochar raised the level of nitrogen, phosphorus, and potassium in soils, relative to a control. Soil microorganisms occupy microhabitats that provide specialized metabolic requirements. Aerobic bacteria present in soil aggregates have been reported to denitrify moist areas of soil peds (Sexstone et al., 1985). Organic matter breakdown rates are higher near the surface of soil aggregates than in the core due to a higher influx of resources. This is due to a reduction in the C C: N ratio, as well as lignin phenol oxidation, resulting from the development of polysaccharides at the surface of soil aggregates (Amelung & Zech, 1996).

Biochar particles in the soil have substantially more oxidized external surfaces than the interior surface or core of soil aggregates (Cheng, 2008; Liang et al., 2006). The adsorption of OM on the surface of biochar particles and C oxidation (Liang et al., 2006) helps to create and maintain the biotic and abiotic reactions mediated by O_2 (Cheng et al., 2008). The preferential oxidation on the particle surface relative to the core enhances the diffusion of O_2 inside biochar particles. Differential redox reactions not only affect the oxidation of organic materials but also transform metal ions. Biochar was reported to enhance alkaline phosphomonoesters and fluorescein diacetate activity relative to a control (Cheng et al., 2008). Jabborova (2020) reported similar results demonstrating that soybean biochar enhances the soil enzymatic activity and biochar application was also previously reported to increase soil enzymatic activity (Bailey et al., 2011; Ma et al., 2019). Some studies have reported an increase in protease, chymotrypsin, trypsin, phosphohydrolase, lipase-esterase, and esterase activity in soils by the application of biochar (Anderson et al., 2011; El Nahhas et al., 2021; Trupiano et al., 2017).

Biochar treatment was reported to significantly enhance urease, invertase, and phosphatase activity in soils (Oladele, 2019) when biochar was applied at a maximum rate (12 t ha^{-1}) at different (0-0.1m) soil depths. Microbial biomass and AMF spores in soils also increased due to the application of biochar, relative to a control. A comparable rise in AMF spores due to biochar application was reported by Hashem et al. (2019). Biochar application has been reported to increase AMF colonization rates in other studies (B. S. Mickan et al., 2016; Vanek & Lehmann, 2015). AMF treatment was reported to enhance alkaline phosphate-monoestrase, microbial biomass, AMF spores, and dehydrogenase activity in soils relative to a control. Other studies reported that biochar increased urease and phosphatase activity, as well as soil microbial biomass (Li & Cai, 2021; Zhaoxiang et al., 2020). Li and Cai (2021) reported that the combined use of AMF and biochar r significantly increased microbial biochemical activity in the soils planted with maize.

Conclusion

This chapter highlights the beneficial effect of biochar and AMF applications on the properties of different soil types, impacting soil biochemistry, microbial activity, AMF colonization, and soil enzyme activity, all of which alters the soil rhizosphere. Biochar has multidimensional benefits when added to soils, including a reduction in greenhouse gas emission (GHGs), increased soil organic matter, soil quality, a reduction in leechable pollutants, the ability to stabilize pesticides, and reclaim heavy metals at different soil depths. The application of AMF to soils in combination with biochar increases the potential benefits of using either compound alone. The combined application of AMF + biochar reduces the leaching of pollutants into groundwater and enhances crop productivity.

The application of new vs. old (weathered) biochar has different advantages and disadvantages on different soil types. Therefore, continued research is needed on this topic, along with the combined use of biochar and AMF. Such studies will provide a better understanding of the mechanisms involved in the improvement of soil topology, acidic or alkaline, as well as enzymatic reactions and their direct and/or indirect effect in different soil types.

A better understanding of the variables that influence the synergistic impact of the combined use of AMF and biochar on biotic and abiotic processes is also needed. Notably, the lack of an effect of biochar on AMF colonization of roots and AMF structure has also been reported, as well as the detrimental impact of biochar on certain soil fungi, such as *Funneliformis mosseae.* It has also been suggested that biochar interferes with the transfer of P from the soil to roots by directly reducing the bioavailability of oxygen in the soil through the absorption of free orthophosphate ions from the soil solution. All non-mycorrhizal species of plants growing in alkaline or acidic soils amended with biochar were reported to have a P deficiency and exhibit symptoms of P deficiency, such as violet leaf coloration. Despite these reported negative impacts, the use of biochar has been shown to enhance root morphology and plant growth. It also has a beneficial impact on enzymatic activity in the soil.

Biochar and AMF applied together exhibited a significant beneficial effect on the growth of spinach, including root morphology, physiological properties, and soil enzymatic activity. We believe that continued research on the use of biochar has the potential to reduce the need for mineral fertilizers. The use of a combination of biochar and AMF as a bio-fertilizer under field conditions was reported to improve spinach plant growth and overall yield.

Research needs to be conducted in different localities in different parts of the world and with different soil types to validate the beneficial impact of biochar and AMF on soil fertility and plant growth and develop fact-based recommendation for its use. Additional trials on other plant species and soil types are also needed to fully understand the benefits of this technology. Isotope labeling studies are also needed to identify the trace nutrients that are available after the application of biochar and AMF. Lastly, carbon flow in soil-plant systems treated with biochar and AMF are needed to fully determine their effect on the physic-chemical properties of soils.

References

Abideen, Z., Koyro, H. W., Huchzermeyer, B., Bilquees, G., & Khan, M. A. (2020). Impact of a biochar or a biochar-compost mixture on water relation, nutrient uptake and photosynthesis of Phragmites karka. *Pedosphere*, *30*(4), 466-477.

Adhikari, S., Gasco, G., Mendez, A., Surapaneni, A., Jegatheesan, V., Shah, K., & Paz-Ferreiro, J. (2019). Influence of pyrolysis parameters on phosphorus fractions of biosolids derived biochar. *Science of the Total Environment*, *695*, 133846.

Ahmad, M., Rajapaksha, A. U., Lim, J. E., Zhang, M., Bolan, N., Mohan, D., Vithanage, M., Lee, S. S., & Ok, Y. S. (2014). Biochar as a sorbent for contaminant management in soil and water: a review. *Chemosphere*, *99*, 19-33.

Alling, V., Hale, S. E., Martinsen, V., Mulder, J., Smebye, A., Breedveld, G. D., & Cornelissen, G. (2014). The role of biochar in retaining nutrients in amended tropical soils. *Journal of Plant Nutrition and Soil Science*, *177*(5), 671-680.

Amelung, W., & Zech, W. (1996). Organic species in ped surface and core fractions along a climosequence in the prairie, North America. *Geoderma*, *74*(3-4), 193-206.

Anderson, C. R., Condron, L. M., Clough, T. J., Fiers, M., Stewart, A., Hill, R. A., & Sherlock, R. R. (2011). Biochar induced soil microbial community change: implications for biogeochemical cycling of carbon, nitrogen and phosphorus. *Pedobiologia*, *54*(5-6), 309-320.

Andrade, L. R., Leal, R. N., Noseda, M., Duarte, M. E. R., Pereira, M. S., Mourão, P. A., Farina, M., & Amado Filho, G. M. (2010). Brown algae overproduce cell wall polysaccharides as a protection mechanism against the heavy metal toxicity. *Marine Pollution Bulletin*, *60*(9), 1482-1488.

Andrade, S., Silveira, A., & Mazzafera, P. (2010). Arbuscular mycorrhiza alters metal uptake and the physiological response of Coffea arabica seedlings to increasing Zn and Cu concentrations in soil. *Science of the Total Environment*, *408*(22), 5381-5391.

Atkinson, C. J., Fitzgerald, J. D., & Hipps, N. A. (2010). Potential mechanisms for achieving agricultural benefits from biochar application to temperate soils: a review. *Plant and Soil*, *337*(1), 1-18.

Bailey, V. L., Fansler, S. J., Smith, J. L., & Bolton Jr, H. (2011). Reconciling apparent variability in effects of biochar amendment on soil enzyme activities by assay optimization. *Soil Biology and Biochemistry*, *43*(2), 296-301.

Biederman, L. A., & Harpole, W. S. (2013). Biochar and its effects on plant productivity and nutrient cycling: a meta-analysis. *GCB Bioenergy*, *5*(2), 202-214.

Blackwell, T. K., Steinbaugh, M. J., Hourihan, J. M., Ewald, C. Y., & Isik, M. (2015). SKN-1/Nrf, stress responses, and aging in Caenorhabditis elegans. *Free Radical Biology and Medicine*, *88*, 290-301.

Bornø, M. L., Müller-Stöver, D. S., & Liu, F. (2018). Contrasting effects of biochar on phosphorus dynamics and bioavailability in different soil types. *Science of the Total Environment*, *627*, 963-974.

Brodowski, S., John, B., Flessa, H., & Amelung, W. (2006). Aggregate-occluded black carbon in soil. *European Journal of Soil Science*, *57*(4), 539-546.

Brussaard, L. (1997). Biodiversity and ecosystem functioning in soil. *Ambio*, 563-570.

Bukovská, P., Bonkowski, M., Konvalinková, T., Beskid, O., Hujslová, M., Püschel, D., Řezáčová, V., Gutiérrez-Núñez, M. S., Gryndler, M., & Jansa, J. (2018). Utilization of organic nitrogen by arbuscular mycorrhizal fungi—is there a specific role for protists and ammonia oxidizers? *Mycorrhiza*, *28*(3), 269-283.

Butnan, S., Deenik, J. L., Toomsan, B., Antal, M. J., & Vityakon, P. (2015). Biochar characteristics and application rates affecting corn growth and properties of soils contrasting in texture and mineralogy. *Geoderma*, *237*, 105-116.

Camenzind, T., Hammer, E. C., Lehmann, J., Solomon, D., Horn, S., Rillig, M. C., & Hempel, S. (2018). Arbuscular mycorrhizal fungal and soil microbial communities in African Dark Earths. *FEMS Microbiology Ecology*, *94*(4), fiy033.

Cao, X., & Harris, W. (2010). Properties of dairy-manure-derived biochar pertinent to its potential use in remediation. *Bioresource Technology*, *101*(14), 5222-5228.

Chang, J., Clay, D. E., Clay, S. A., Chintala, R., Miller, J. M., & Schumacher, T. (2016). Biochar reduced nitrous oxide and carbon dioxide emissions from soil with different water and temperature cycles. *Agronomy Journal*, *108*(6), 2214.

Cheng, H., Jones, D. L., Hill, P., & Bastami, M. S. (2017). Biochar concomitantly increases simazine sorption in sandy loam soil and lowers its dissipation. *Archives of Agronomy and Soil Science*, *63*(8), 1082-1092.

Cheng, L., Booker, F. L., Tu, C., Burkey, K. O., Zhou, L., Shew, H. D., Rufty, T. W., & Hu, S. (2012). Arbuscular mycorrhizal fungi increase organic carbon decomposition under elevated CO_2. *Science*, *337*(6098), 1084-1087.

Cheng, P., Burr, G. S., Zhou, W., Chen, N., Hou, Y., Du, H., Fu, Y., & Lu, X. (2020). The deficiency of organic matter 14C dating in Chinese Loess-paleosol sample. *Quaternary Geochronology*, *56*, 101051.

Cheng, Q. (2008). Non-linear theory and power-law models for information integration and mineral resources quantitative assessments. *Mathematical Geosciences*, *40*(5), 503-532.

Cosme, M., & Wurst, S. (2013). Interactions between arbuscular mycorrhizal fungi, rhizobacteria, soil phosphorus and plant cytokinin deficiency change the root morphology, yield and quality of tobacco. *Soil Biology and Biochemistry*, *57*, 436-443.

de Figueiredo, C. C., Farias, W. M., Coser, T. R., de Paula, A. M., Da Silva, M. R. S., & Paz-Ferreiro, J. (2019). Sewage sludge biochar alters root colonization of mycorrhizal fungi in a soil cultivated with corn. *European Journal of Soil Biology*, *93*, 103092.

Efthymiou, A., Jensen, B., & Jakobsen, I. (2018). The roles of mycorrhiza and Penicillium inoculants in phosphorus uptake by biochar-amended wheat. *Soil Biology and Biochemistry*, *127*, 168-177.

El Nahhas, N., AlKahtani, M. D., Abdelaal, K. A., Al Husnain, L., AlGwaiz, H. I., Hafez, Y. M., Attia, K. A., El-Esawi, M. A., Ibrahim, M. F., & Elkelish, A. (2021). Biochar and jasmonic acid application attenuates antioxidative systems and improves growth, physiology, nutrient uptake and productivity of faba bean (Vicia faba L.) irrigated with saline water. *Plant Physiology and Biochemistry*, *166*, 807-817.

Farzaneh, M., Vierheilig, H., Lössl, A., & Kaul, H. (2011). Arbuscular mycorrhiza enhances nutrient uptake in chickpea. *Plant, Soil and Environment*, *57*(10), 465-470.

Göhre, V., & Paszkowski, U. (2006). Contribution of the arbuscular mycorrhizal symbiosis to heavy metal phytoremediation. *Planta*, *223*(6), 1115-1122.

Gryndler, M., Šmilauer, P., Püschel, D., Bukovská, P., Hršelová, H., Hujslová, M., Gryndlerová, H., Beskid, O., Konvalinková, T., & Jansa, J. (2018). Appropriate nonmycorrhizal controls in arbuscular mycorrhiza research: a microbiome perspective. *Mycorrhiza*, *28*(5), 435-450.

Gul, S., & Whalen, J. K. (2016). Biochemical cycling of nitrogen and phosphorus in biochar-amended soils. *Soil Biology and Biochemistry*, *103*, 1-15.

Gundale, M. J., & DeLuca, T. H. (2006). Temperature and source material influence ecological attributes of ponderosa pine and Douglas-fir charcoal. *Forest Ecology and Management*, *231*(1-3), 86-93.

Hammer, E. C., Balogh-Brunstad, Z., Jakobsen, I., Olsson, P. A., Stipp, S. L., & Rillig, M. C. (2014). A mycorrhizal fungus grows on biochar and captures phosphorus from its surfaces. *Soil Biology and Biochemistry*, *77*, 252-260.

Hammer, E. C., Forstreuter, M., Rillig, M. C., & Kohler, J. (2015). Biochar increases arbuscular mycorrhizal plant growth enhancement and ameliorates salinity stress. *Applied Soil Ecology*, *96*, 114-121.

Hashem, A., Kumar, A., Al-Dbass, A. M., Alqarawi, A. A., Al-Arjani, A.-B. F., Singh, G., Farooq, M., & Abd_Allah, E. F. (2019). Arbuscular mycorrhizal fungi and biochar improves drought tolerance in chickpea. *Saudi Journal of Biological Sciences*, *26*(3), 614-624.

Hazard, C., Gosling, P., Van Der Gast, C. J., Mitchell, D. T., Doohan, F. M., & Bending, G. D. (2013). The role of local environment and geographical distance in determining community composition of arbuscular mycorrhizal fungi at the landscape scale. *The ISME Journal*, *7*(3), 498-508.

Hodge, A., & Fitter, A. H. (2010). Substantial nitrogen acquisition by arbuscular mycorrhizal fungi from organic material has implications for N cycling. *Proceedings of the National Academy of Sciences*, *107*(31), 13754-13759.

Hodge, A., & Storer, K. (2015). Arbuscular mycorrhiza and nitrogen: implications for individual plants through to ecosystems. *Plant and Soil*, *386*(1), 1-19.

Hu, Y., Liu, X., Bai, J., Shih, K., Zeng, E. Y., & Cheng, H. (2013). Assessing heavy metal pollution in the surface soils of a region that had undergone three decades of intense industrialization and urbanization. *Environmental Science and Pollution Research*, *20*(9), 6150-6159.

Ibrahim, M. M., Tong, C., Hu, K., Zhou, B., Xing, S., & Mao, Y. (2020). Biochar-fertilizer interaction modifies N-sorption, enzyme activities and microbial functional abundance regulating nitrogen retention in rhizosphere soil. *Science of the Total Environment*, *739*, 140065.

Jabborova, A. (2020). Phraseologisms Expressing the Emotional State of a Person (Face). *Mental Enlightenment Scientific-Methodological Journal*, *2020*(2), 45-56.

Jabborova, D., Wirth, S., Kannepalli, A., Narimanov, A., Desouky, S., Davranov, K., Sayyed, R. Z., El Enshasy, H., Malek, R. A., & Syed, A. (2020). Co-inoculation of rhizobacteria and biochar application improves growth and nutrientsin soybean and enriches soil nutrients and enzymes. *Agronomy*, *10*(8), 1142.

Jansa, J., Erb, A., Oberholzer, H. R., Šmilauer, P., & Egli, S. (2014). Soil and geography are more important determinants of indigenous arbuscular mycorrhizal communities than management practices in Swiss agricultural soils. *Molecular Ecology*, *23*(8), 2118-2135.

Jeffery, S., Abalos, D., Prodana, M., Bastos, A. C., Van Groenigen, J. W., Hungate, B. A., & Verheijen, F. (2017). Biochar boosts tropical but not temperate crop yields. *Environmental Research Letters*, *12*(5), 053001.

Jeffery, S., Verheijen, F. G., van der Velde, M., & Bastos, A. C. (2011). A quantitative review of the effects of biochar application to soils on crop productivity using meta-analysis. *Agriculture, Ecosystems & Environment*, *144*(1), 175-187.

Johnson, N. C., Gehring, C., & Jansa, J. (2016). *Mycorrhizal Mediation of Soil: Fertility, Structure, and Carbon Storage*. Elsevier.

Jones, D., Edwards-Jones, G., & Murphy, D. (2011). Biochar mediated alterations in herbicide breakdown and leaching in soil. *Soil Biology and Biochemistry*, *43*(4), 804-813.

Kang, F., Yang, B., Yang, X., Wang, L., Guo, J., Sun, W., Zhang, Q., & Zhang, T. (2020). Arbuscular mycorrhizal fungi alleviate the negative effect of nitrogen deposition on ecosystem functions in meadow grassland. *Land Degradation & Development*, *31*(6), 748-759.

Khorram, M. S., Wang, Y., Jin, X., Fang, H., & Yu, Y. (2015). Reduced mobility of fomesafen through enhanced adsorption in biochar-amended soil. *Environmental Toxicology and Chemistry*, *34*(6), 1258-1266.

Kloss, S., Zehetner, F., Dellantonio, A., Hamid, R., Ottner, F., Liedtke, V., Schwanninger, M., Gerzabek, M. H., & Soja, G. (2012). Characterization of slow pyrolysis biochars: effects of feedstocks and pyrolysis temperature on biochar properties. *Journal of Environmental Quality*, *41*(4), 990-1000.

Koide, R. (2017). Biochar—Arbuscular mycorrhiza interaction in temperate soils. In *Mycorrhizal Mediation of Soil* (pp. 461-477). Elsevier.

Koide, R., & Kabir, Z. (2000). Extraradical hyphae of the mycorrhizal fungus Glomus intraradices can hydrolyse organic phosphate. *New Phytologist*, *148*(3), 511-517.

Kuppusamy, S., Thavamani, P., Megharaj, M., Venkateswarlu, K., & Naidu, R. (2016). Agronomic and remedial benefits and risks of applying biochar to soil: current knowledge and future research directions. *Environment International*, *87*, 1-12.

Labarrere, C. A., Woods, J., Hardin, J., Campana, G., Ortiz, M., Jaeger, B., Reichart, B., Bonnin, J., Currin, A., & Cosgrove, S. (2011). Early prediction of cardiac allograft vasculopathy and heart transplant failure. *American Journal of Transplantation*, *11*(3), 528-535.

Laird, D. A., Fleming, P., Davis, D. D., Horton, R., Wang, B., & Karlen, D. L. (2010). Impact of biochar amendments on the quality of a typical Midwestern agricultural soil. *Geoderma*, *158*(3-4), 443-449.

Lehmann, J., Gaunt, J., & Rondon, M. (2006). Bio-char sequestration in terrestrial ecosystems–a review. *Mitigation and Adaptation Strategies for Global Change*, *11*(2), 403-427.

Lehmann, J., Rillig, M. C., Thies, J., Masiello, C. A., Hockaday, W. C., & Crowley, D. (2011). Biochar effects on soil biota–a review. *Soil Biology and Biochemistry*, *43*(9), 1812-1836.

Li, M., & Cai, L. (2021). Biochar and Arbuscular Mycorrhizal Fungi Play Different Roles in Enabling Maize to Uptake Phosphorus. *Sustainability*, *13*(6), 3244.

Liang, B., Lehmann, J., Solomon, D., Kinyangi, J., Grossman, J., O'Neill, B., Skjemstad, J. O., Thies, J., Luizão, F. J., & Petersen, J. (2006). Black carbon increases cation exchange capacity in soils. *Soil Science Society of America Journal*, *70*(5), 1719-1730.

Liu, L., Li, W., Song, W., & Guo, M. (2018). Remediation techniques for heavy metal-contaminated soils: Principles and applicability. *Science of the Total Environment*, *633*, 206-219.

Luo, S., Wang, S., Tian, L., Li, S., Li, X., Shen, Y., & Tian, C. (2017). Long-term biochar application influences soil microbial community and its potential roles in semiarid farmland. *Applied Soil Ecology*, *117*, 10-15.

Ma, H., Egamberdieva, D., Wirth, S., & Bellingrath-Kimura, S. D. (2019). Effect of biochar and irrigation on soybean-rhizobium symbiotic performance and soil enzymatic activity in field rhizosphere. *Agronomy*, *9*(10), 626.

Mao, J. D., Johnson, R., Lehmann, J., Olk, D., Neves, E., Thompson, M., & Schmidt-Rohr, K. (2012). Abundant and stable char residues in soils: implications for soil fertility and carbon sequestration. *Environmental Science & Technology*, *46*(17), 9571-9576.

Maraseni, T. N. (2010). Biochar: maximising the benefits. *International Journal of Environmental Studies*, *67*(3), 319-327.

Marín-Benito, J., Brown, C., Herrero-Hernández, E., Arienzo, M., Sánchez-Martín, M., & Rodríguez-Cruz, M. (2013). Use of raw or incubated organic wastes as amendments in reducing pesticide leaching through soil columns. *Science of the Total Environment*, *463*, 589-599.

Mickan, B. S., Abbott, L. K., Stefanova, K., & Solaiman, Z. M. (2016). Interactions between biochar and mycorrhizal fungi in a water-stressed agricultural soil. *Mycorrhiza*, *26*(6), 565-574.

Mickan, M., Helmersson, U., Rinnert, H., Ghanbaja, J., Muller, D., & Horwat, D. (2016). Room temperature deposition of homogeneous, highly transparent and conductive Al-doped ZnO films by reactive high power impulse magnetron sputtering. *Solar Energy Materials and Solar Cells*, *157*, 742-749.

Nelly, N., Syahrawati, M., Hamid, H., Habazar, T., & Gusnia, D. N. (2019). Diversity and characterization of entomopathogenic fungi from rhizosphere of maize plants as potential biological control agents. *Biodiversitas Journal of Biological Diversity*, *20*(5).

Nishida, H., Tanaka, S., Handa, Y., Ito, M., Sakamoto, Y., Matsunaga, S., Betsuyaku, S., Miura, K., Soyano, T., & Kawaguchi, M. (2018). A NIN-LIKE PROTEIN mediates nitrate-induced control of root nodule symbiosis in Lotus japonicus. *Nature Communications*, *9*(1), 1-14.

Oladele, S. O. (2019). Effect of biochar amendment on soil enzymatic activities, carboxylate secretions and upland rice performance in a sandy clay loam Alfisol of Southwest Nigeria. *Scientific African*, *4*, e00107.

Paymaneh, Z., Gryndler, M., Konvalinková, T., Benada, O., Borovička, J., Bukovská, P., Püschel, D., Řezáčová, V., Sarcheshmehpour, M., & Jansa, J. (2018). Soil matrix determines the outcome of interaction between mycorrhizal symbiosis and biochar for Andropogon gerardii growth and nutrition. *Frontiers in Microbiology*, 2862.

Prendergast-Miller, M., Duvall, M., & Sohi, S. (2014). Biochar–root interactions are mediated by biochar nutrient content and impacts on soil nutrient availability. *European Journal of Soil Science*, *65*(1), 173-185.

Qayyum, M. F., Haider, G., Raza, M. A., Mohamed, A. K. S., Rizwan, M., El-Sheikh, M. A., Alyemeni, M. N., & Ali, S. (2020). Straw-based biochar mediated potassium availability and increased growth and yield of cotton (Gossypium hirsutum L.). *Journal of Saudi Chemical Society*, *24*(12), 963-973.

Ren, X., Wang, F., Zhang, P., Guo, J., & Sun, H. (2018). Aging effect of minerals on biochar properties and sorption capacities for atrazine and phenanthrene. *Chemosphere*, *206*, 51-58.

Richardson, A. E. (2001). Prospects for using soil microorganisms to improve the acquisition of phosphorus by plants. *Functional Plant Biology*, *28*(9), 897-906.

Roberts, K. G., Gloy, B. A., Joseph, S., Scott, N. R., & Lehmann, J. (2010). Life cycle assessment of biochar systems: estimating the energetic, economic, and climate change potential. *Environmental Science & Technology*, *44*(2), 827-833.

Sarthou, J. P., Badoz, A., Vaissière, B., Chevallier, A., & Rusch, A. (2014). Local more than landscape parameters structure natural enemy communities during their overwintering in semi-natural habitats. *Agriculture, Ecosystems & Environment*, *194*, 17-28.

Saxena, J., Rana, G., & Pandey, M. (2013). Impact of addition of biochar along with Bacillus sp. on growth and yield of French beans. *Scientia Horticulturae*, *162*, 351-356.

Schulzke, T. (2019). Biomass gasification: conversion of forest residues into heat, electricity and base chemicals. *Chemical Papers*, *73*(8), 1833-1852.

Ścisłowska, M., Włodarczyk, R., Kobyłecki, R., & Bis, Z. (2015). Biochar to improve the quality and productivity of soils. *Journal of Ecological Engineering*, *16*(3).

Sexstone, A. J., Revsbech, N. P., Parkin, T. B., & Tiedje, J. M. (1985). Direct measurement of oxygen profiles and denitrification rates in soil aggregates. *Soil Science Society of America Journal*, 49 (3).

Shen, Q., Hedley, M., Camps Arbestain, M., & Kirschbaum, M. (2016). Can biochar increase the bioavailability of phosphorus? *Journal of Soil Science and Plant Nutrition*, *16*(2), 268-286.

Smith, S. E., Facelli, E., Pope, S., & Andrew Smith, F. (2010). Plant performance in stressful environments: interpreting new and established knowledge of the roles of arbuscular mycorrhizas. *Plant and Soil*, *326*(1), 3-20.

Smith, S. E., & Smith, F. A. (2012). Fresh perspectives on the roles of arbuscular mycorrhizal fungi in plant nutrition and growth. *Mycologia*, *104*(1), 1-13.

Smith, S. E., Smith, F. A., & Jakobsen, I. (2004). Functional diversity in arbuscular mycorrhizal (AM) symbioses: the contribution of the mycorrhizal P uptake pathway is not correlated with mycorrhizal responses in growth or total P uptake. *New Phytologist*, *162*(2), 511-524.

Söderberg, C., & Eckerberg, K. (2013). Rising policy conflicts in Europe over bioenergy and forestry. *Forest Policy and Economics*, *33*, 112-119.

Solaiman, Z. M., Blackwell, P., Abbott, L. K., & Storer, P. (2010). Direct and residual effect of biochar application on mycorrhizal root colonisation, growth and nutrition of wheat. *Soil Research*, *48*(7), 546-554.

Song, D., Chen, L., Zhang, S., Zheng, Q., Ullah, S., Zhou, W., & Wang, X. (2020). Combined biochar and nitrogen fertilizer change soil enzyme and microbial activities in a 2-year field trial. *European Journal of Soil Biology*, *99*, 103212.

Song, Y., Zhang, X., Ma, B., Chang, S. X., & Gong, J. (2014). Biochar addition affected the dynamics of ammonia oxidizers and nitrification in microcosms of a coastal alkaline soil. *Biology and Fertility of Soils*, *50*(2), 321-332.

Spagnoletti, F., Carmona, M., Gómez, N. E. T., Chiocchio, V., & Lavado, R. S. (2017). Arbuscular mycorrhiza reduces the negative effects of M. phaseolina on soybean plants in arsenic-contaminated soils. *Applied Soil Ecology*, *121*, 41-47.

Spokas, K. A., Cantrell, K. B., Novak, J. M., Archer, D. W., Ippolito, J. A., Collins, H. P., Boateng, A. A., Lima, I. M., Lamb, M. C., & McAloon, A. J. (2012). Biochar: a synthesis of its agronomic impact beyond carbon sequestration. *Journal of Environmental Quality*, *41*(4), 973-989.

Srinivasan, P., & Sarmah, A. K. (2015). Characterisation of agricultural waste-derived biochars and their sorption potential for sulfamethoxazole in pasture soil: a spectroscopic investigation. *Science of the Total Environment*, *502*, 471-480.

Steinbeiss, S., Gleixner, G., & Antonietti, M. (2009). Effect of biochar amendment on soil carbon balance and soil microbial activity. *Soil Biology and Biochemistry*, *41*(6), 1301-1310.

Steiner, K., Novotny, R., Werz, D. B., Zarschler, K., Seeberger, P. H., Hofinger, A., Kosma, P., Schäffer, C., & Messner, P. (2008). Molecular basis of S-layer glycoprotein glycan biosynthesis in Geobacillus stearothermophilus. *Journal of Biological Chemistry*, *283*(30), 21120-21133.

Tian, J., Wang, J., Dippold, M., Gao, Y., Blagodatskaya, E., & Kuzyakov, Y. (2016). Biochar affects soil organic matter cycling and microbial functions but does not alter microbial community structure in a paddy soil. *Science of the Total Environment*, *556*, 89-97.

Tisserant, E., Malbreil, M., Kuo, A., Kohler, A., Symeonidi, A., Balestrini, R., Charron, P., Duensing, N., dit Frey, N. F., & Gianinazzi-Pearson, V. (2013). Genome of an arbuscular mycorrhizal fungus provides insight into the oldest plant symbiosis. *Proceedings of the National Academy of Sciences*, *110*(50), 20117-20122.

Tomè, E., Tagliavini, M., & Scandellari, F. (2015). Recently fixed carbon allocation in strawberry plants and concurrent inorganic nitrogen uptake through arbuscular mycorrhizal fungi. *Journal of Plant Physiology*, *179*, 83-89.

Treseder, K. K., & Allen, M. F. (2002). Direct nitrogen and phosphorus limitation of arbuscular mycorrhizal fungi: a model and field test. *New Phytologist*, *155*(3), 507-515.

Trupiano, D., Cocozza, C., Baronti, S., Amendola, C., Vaccari, F. P., Lustrato, G., Di Lonardo, S., Fantasma, F., Tognetti, R., & Scippa, G. S. (2017). The effects of biochar and its combination with compost on lettuce (Lactuca sativa L.) growth, soil properties, and soil microbial activity and abundance. *International Journal of Agronomy*, *2017*.

van der Heijden, M. G., Martin, F. M., Selosse, M. A., & Sanders, I. R. (2015). Mycorrhizal ecology and evolution: the past, the present, and the future. *New Phytologist*, *205*(4), 1406-1423.

Vanek, S. J., & Lehmann, J. (2015). Phosphorus availability to beans via interactions between mycorrhizas and biochar. *Plant and Soil*, *395*(1), 105-123.

Vassilev, N., Martos, E., Mendes, G., Martos, V., & Vassileva, M. (2013). Biochar of animal origin: a sustainable solution to the global problem of high-grade rock phosphate scarcity? *Journal of the Science of Food and Agriculture*, *93*(8), 1799-1804.

Vejvodová, K., Vaněk, A., Mihaljevič, M., Ettler, V., Trubač, J., Vaňková, M., Drahota, P., Vokurkova, P., Penížek, V., & Zadorova, T. (2020). Thallium isotopic fractionation in soil: the key controls. *Environmental Pollution*, *265*, 114822.

Wang, Y., Yin, R., & Liu, R. (2014). Characterization of biochar from fast pyrolysis and its effect on chemical properties of the tea garden soil. *Journal of Analytical and Applied Pyrolysis*, *110*, 375-381.

Warnock, D. D., Lehmann, J., Kuyper, T. W., & Rillig, M. C. (2007). Mycorrhizal responses to biochar in soil–concepts and mechanisms. *Plant and Soil*, *300*(1), 9-20.

Warnock, D. D., Mummey, D. L., McBride, B., Major, J., Lehmann, J., & Rillig, M. C. (2010). Influences of non-herbaceous biochar on arbuscular mycorrhizal fungal abundances in roots and soils: results from growth-chamber and field experiments. *Applied Soil Ecology*, *46*(3), 450-456.

Williams, M., Martin, S., & Kookana, R. S. (2015). Sorption and plant uptake of pharmaceuticals from an artificially contaminated soil amended with biochars. *Plant and Soil*, *395*(1), 75-86.

Yamato, M., Okimori, Y., Wibowo, I. F., Anshori, S., & Ogawa, M. (2006). Effects of the application of charred bark of Acacia mangium on the yield of maize, cowpea and peanut, and soil chemical properties in South Sumatra, Indonesia. *Soil Science and Plant Nutrition*, *52*(4), 489-495.

Yu, X. Y., Ying, G. G., & Kookana, R. S. (2009). Reduced plant uptake of pesticides with biochar additions to soil. *Chemosphere*, *76*(5), 665-671.

Zamin, M., Fahad, S., Khattak, A. M., Adnan, M., Wahid, F., Raza, A., Wang, D., Saud, S., Noor, M., & Bakhat, H. F. (2020). Developing the first halophytic turfgrasses for the urban landscape from native Arabian desert grass. *Environmental Science and Pollution Research*, *27*(32), 39702-39716.

Zhang, D., Zhang, C., Tang, X., Li, H., Zhang, F., Rengel, Z., Whalley, W. R., Davies, W. J., & Shen, J. (2016). Increased soil phosphorus availability induced by faba bean root exudation stimulates root growth and phosphorus uptake in neighbouring maize. *New Phytologist*, *209*(2), 823-831.

Zhang, L., Jing, Y., Xiang, Y., Zhang, R., & Lu, H. (2018). Responses of soil microbial community structure changes and activities to biochar addition: a meta-analysis. *Science of the Total Environment*, *643*, 926-935.

Zhang, L., Xu, M., Liu, Y., Zhang, F., Hodge, A., & Feng, G. (2016). Carbon and phosphorus exchange may enable cooperation between an arbuscular mycorrhizal fungus and a phosphate-solubilizing bacterium. *New Phytologist*, *210*(3), 1022-1032.

Zhang, Z., Bian, L., Sun, X., Luo, Z., Xin, Z., Luo, F., & Chen, Z. (2015). Electrophysiological and behavioural responses of the tea geometrid Ectropis obliqua (Lepidoptera: Geometridae) to volatiles from a non-host plant, rosemary, Rosmarinus officinalis (Lamiaceae). *Pest Management Science*, *71*(1), 96-104.

Zhao, H., Yu, L., Yu, M., Afzal, M., Dai, Z., Brookes, P., & Xu, J. (2020). Nitrogen combined with biochar changed the feedback mechanism between soil nitrification and Cd availability in an acidic soil. *Journal of Hazardous Materials*, *390*, 121631.

Zhao, Z., Chen, L., & Xiao, Y. (2021). The combined use of arbuscular mycorrhizal fungi, biochar, and nitrogen fertilizer is most beneficial to cultivating Cichorium intybus L. in Cd-contaminated soil. *Ecotoxicology and Environmental Safety*, *217*, 112154.

Zhaoxiang, W., Huihu, L., Qiaoli, L., Changyan, Y., & Faxin, Y. (2020). Application of bio-organic fertilizer, not biochar, in degraded red soil improves soil nutrients and plant growth. *Rhizosphere*, *16*, 100264.

Zheng, H., Wang, X., Chen, L., Wang, Z., Xia, Y., Zhang, Y., Wang, H., Luo, X., & Xing, B. (2018). Enhanced growth of halophyte plants in biochar-amended coastal soil: roles of nutrient availability and rhizosphere microbial modulation. *Plant, Cell & Environment*, *41*(3), 517-532.

Zwetsloot, M. J., Lehmann, J., Bauerle, T., Vanek, S., Hestrin, R., & Nigussie, A. (2016). Phosphorus availability from bone char in a P-fixing soil influenced by root-mycorrhizae-biochar interactions. *Plant and Soil*, *408*(1), 95-105.

Chapter 3

Plant-Soil Feedback Regulates the Invasive Competition of Alien Plants through Arbuscular Mycorrhizal Fungi

Kai-Ping Shen
Yun Guo
and Yue-Jun He*
College of Forestry, Research Center of Forest Ecology,
Guizhou University, Guiyang, China

Abstract

Plant invasion has become a global environmental, economic and social problem, posing a serious threat to the balance and stability of ecosystems, even to human health and socio-economic development. Arbuscular mycorrhizal (AM) fungi can regulate plant growth, nutrient acquisition, and the competitive relationship between alien plants and native plants. Therefore, AM fungi play a crucial role in the invasion of alien plants, exerting different feedback effects on their growth and competition. Studying the feedback effects of AM fungi on plant invasion has become a major part of mycorrhizal and invasion ecology research.

This chapter focuses on the feedback effects and main mechanisms of action of AM fungi in competition between invasive and native plants from the perspective of AM fungi regulating plant nutrient uptake and growth to influence interspecific competition of plants. The aim is to offer a thorough understanding of the mechanisms of action of AM fungi

* Corresponding Author's Email: hyj1358@163.com.

In: New Research on Mycorrhizal Fungus
Editors: Qiang-Sheng Wu, Ying-Ning Zou, Yue-Jun He et al.
ISBN: 979-8-88697-637-3

in the invasive plant-soil feedback process, which can provide a theoretical basis for the biological control of harmful alien plants and the restoration of damaged or fragile ecosystems and the maintenance of ecological security.

Keywords: arbuscular mycorrhizal fungi, invasion, competition, plant-soil feedback

1. Introduction

Biological invasion is a process whereby an alien species invade a new habitat for natural or anthropogenic reasons and settle there (Lodge 1993, van Kleunen et al. 2015). A successful plant invasion must undergo four stages: introduction, colonization, population establishment, and dispersal (Theoharides and Dukes 2007). The superiority of alien species over native species can cause native species to be eliminated in competition (Szymura and Szymura 2016), resulting in a stable mono-optimal community (Dukes and Mooney 1999), and adversely affecting the biodiversity of the invaded habitat, which causes a severe problem to the quality of the global ecosystem, economic interests and human health (Pejchar and Mooney 2009, Ehrenfeld 2010). Therefore, biological invasion as a threat to biodiversity has attracted widespread global attention and has become one of the hotspots for global change research (Ehrenfeld 2010). Hence, exploring the mechanisms of a successful invasion of alien plants has been an essential theme in ecology and evolution during recent decades and is also a prerequisite for invasion prevention and control.

Plant-soil feedback (PSF) is an essential mechanism in ecosystems, and it refers to the effect of plants on the biological and abiotic characteristics of rhizosphere soil, which in turn affects the behavior of plants themselves or other species (Ehrenfeld et al. 2005). It includes two main aspects: plants change the soil environment and this changed soil environment, in turn, affects plants themselves or other species. Therefore, the interaction between plants and soil is considered to be the fundamental driver of soil origin (Wardle et al. 2004) and can predict the evolutionary direction of terrestrial plants (Tamme et al. 2010, Simberloff et al. 2013, Van der Putten et al. 2013, Kroon et al. 2015). To some extent, the PSF can explain how exotic plants can successfully invade. The PSF has profound implications for vegetation function regulation of ecosystems (Van der Putten et al. 2013) and the species coexistence of plant communities (Eppinga et al. 2006, Mack and Bever 2014); of course, these coexisting species may also include invasive and native plants. Alien plant invasion affects the growth of native plants through competition (Brown et al. 2002, Cuda et al. 2015), and competitive intensity is key to a successful invasion of alien plants. Additionally, alien plants can also change the soil physicochemical properties of invasive habitats (Weidenhamer and Callaway 2010, Van der Putten et al. 2013), as well as the soil microbial composition and diversity (Mangla et al. 2008), with the altered soil in turn affecting the invasive plants themselves and the growth of native plants. Therefore, alien plant invasion is the process of plant-soil feedback regulation mediated by invasive and native plants.

Meanwhile, soil microorganisms are likely to play an essential role in the PSF of alien and native plant interactions, such as arbuscular mycorrhizal (AM) fungi, as the soil functional microbes can form the mutualistic symbiont with more than two-thirds of terrestrial plant roots (Smith and Read 2008). AM fungi can promote plant growth and nutrient uptake (Hodge and Fitter 2010), improve plant stress tolerance (He and Zhong 2011, He 2019), alter competitive relationships between individual plants (e.g., alien plant and native plant) (Shen et al. 2020, Guo et al. 2021), and further affect species diversity and community composition (Neuenkamp et al. 2019). Therefore, the chapter simply outlined the regulation of AM fungi on competition between invasive and native plants and described an important mechanism as an explanation for plant invasion, i.e., PSF. Meanwhile, we point out some shortcomings of the current research and make some suggestions for future research with the aim of drawing on relevant research and prevention practices for carrying out the role of AM fungi in alien plant invasion.

2. Arbuscular Mycorrhizal Fungi Regulate Competition between Invasive and Native Plants in Nutritional Physiology

2.1. Nutrient Competition of Invasive and Native Plants Regulated by Arbuscular Mycorrhizal Networks

AM fungi commonly have a positive effect on the growth of the host plant because AM fungi extend their absorption network to nutritional areas that the root system cannot reach, which expands the soil area for the host plant to be touched (Smith et al. 2011). Additionally, mycelium is more refined than the root system and can pass through smaller pores and take up more nutrients, thus promoting the growth of the host plant (Diagne et al. 2020, Stock et al. 2021). Especially, AM fungi can promote the growth of host plant by expanding the extraradical mycelium in the soil to form a vast common mycorrhizal network (CMN), resulting in an expanded root uptake area and increased root viability, which allows the plant to absorb significantly more water and nutrients (e.g., N and P) (Kiers et al. 2011). AM fungi can form symbioses by coupling with most of the plant root systems due to their non-specific characteristics (Scheublin et al. 2007). However, mycorrhizal symbiosis involves two host plants interacting with multiple host plants simultaneously, meaning a single plant host is colonized by multiple fungal species and fungal individuals interact with multiple host plants, interconnected through a CMN (Fellbaum et al. 2014). Therefore, the differences in the photosynthetic products provided by different host plants lead to differential resource allocation among host plants through CMN, thus affecting the competitive relationships among plant species (Simard et al. 2003, Merrild et al. 2013, Schott et al. 2016, He et al. 2019).

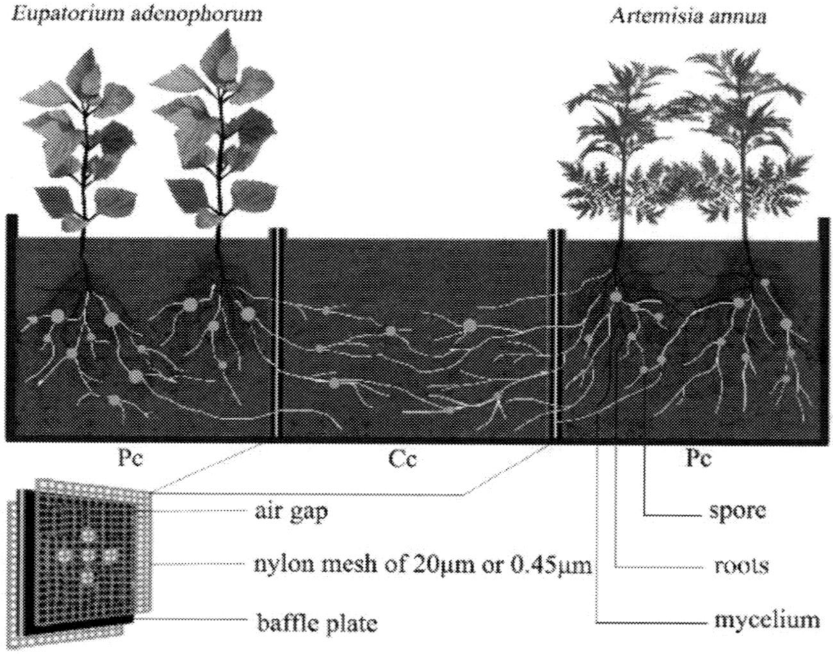

Figure 1. The compartmentalized microcosm growing device between invasive plants and native plants (Shen et al. 2020).

Recent research demonstrated that CMN could mediate competition between invasive and native plants for resources (Workman and Cruzan 2016). The CMN preferentially distributes mineral nutrients to the roots of the invasive plants, thus facilitating the invasion of the alien plants when invasive plants are connected to native plants through the CMN (Awaydul et al. 2019, Řezáčová et al. 2020). For example, Shen et al. (2020) and Xia et al. (2020) found that the mycorrhizal network transferred more nutrient resources to the invasive plant *Eupatorium adenophorum* when it competed with the native plant *Artemisia annua* through a connected mycorrhizal network giving the invasive plant a more developed root system and increasing its belowground competitive advantage over the native plant, thus promoting greater growth and nutrient acquisition advantage of the alien plant versus the native plant (Figure 1).

Additionally, Yu and He (2021) also found that the invasiveness of plants can be strengthened by AM fungi via asymmetrical amino acid competition. Therefore, there is widespread acceptance that AM fungi inhibit the growth of native plants by interfering with their symbiotic relationships (Wilson et al. 2012). However, AM fungi also could potentially assist local plants in defending against alien plants under certain conditions (Cheng et al. 2019). The mycorrhizal symbiosis resistance hypothesis stated that mycorrhizal symbiosis could assist local plants in countering invasion by alien plants (Kisa et al. 2007). Moreover, the influence of AM fungi on the competitive relationship between alien and native plants is regulated by a variety of factors, such as soil fertility. Johnson et al. (2015) found that AM fungi were more likely to promote host plant growth in P-restricted systems and inhibit host plant growth in N-restricted systems. Thus, the mechanisms behind this preference for invasive plants by CMN may be related to plant characteristics, environment, neighboring plants, etc. (Yang et al. 2014, Molina and Horton 2015, Rahman et al. 2019). Therefore, how differences in mycorrhizal preferences for host plants affect resource use and transfer, and in turn alter the competitiveness of invasive plants, still needs to be explored in depth.

2.2. Arbuscular Mycorrhizal Fungi Regulate Plant Growth and Competition by Affecting the Physiological Characteristics of the Invasive Plant

There is a 'trade-off' between the host plant and the AM fungi (Smith and Read 2008). The plant offers the carbon source from photosynthesis to mycorrhizal fungi, which provide the carbon skeleton and energy for mycelial growth and spore development, and the AM fungi provide the nutrients required for the host plant's growth and development (Smith and Read 2008, Smith et al. 2011). Additionally, AM fungi can influence the physiological metabolism of plants, thereby enhancing their ability to tolerate biotic and abiotic stresses. How AM fungi affected the photosynthetic characteristics of *Broussonetia papyrifera* seedlings in karst habitats was researched by He et al. (2008), who discovered that AM fungi significantly increased the net photosynthesis rate (Pn), transpiration rate (E), stomatal conductance (Sc), water use efficiency (WUE), and the content of chlorophyll a (Chl a) and chlorophyll b (Chl b) of *Broussonetia papyrifera* seedlings. According to He and Zhong (2011), under different water stress conditions, AM fungi increased the concentration of soluble sugar, soluble protein, and proline and reduced the concentration of malondialdehyde for *Cinnamomum camphora* seedlings. This research indicated that the AM fungi-plant symbiosis regulates these physiological mechanisms in order to enhance the plant's ability to adapt to its environment.

Further, AM fungi can alter the morphology, physiology, and competitive ability of host plants (Bray et al. 2003, Wu et al. 2018), including invasive plants. AM fungi increased the Pn of native *S. viridis* in monoculture conditions while reducing the Pn when competing with invasive *F. bidentis* (Zhang et al. 2017). The greater growth for alien plants than native plants may inhibit photosynthesis in native plants (Huang et al. 2020), resulting in native plants producing less photosynthetic products, which may not be sufficient to support the resources they need to exchange with AM fungi, and therefore having lower growth and nutrient acquisition capacity (Fellbaum et al. 2012). However, a competitive experiment using a sympatric native plant with an invasive plant showed invasive plants performed worst (as indicated by reduced Chl content) in invaded range soil against conspecifics (Workman 2014).

Therefore, there is debate about AM fungi affect the competitive relationship between invasion and native plants by altering plant physiological characteristics. Meanwhile, there are still few studies on the regulation of invasion by AM fungi in physiological competition with native plants, especially for field experiments. Future studies should consider these aspects to understand better the physiological mechanisms of the invasion of alien plants.

3. The Plant-Soil Feedback Regulating Plant Invasion through Arbuscular Mycorrhizal Fungi

3.1. The Positive and Negative Regulation of Plant-Soil Feedback in the Ecological Invasion

The concept of plant-soil feedback (PSF) was introduced by Bever et al. (1997), whereby plants affect the soil microbial community, and the microbial community responds to the plants. It was later added to by Van der Putten et al. (2013) and generalized into direct and indirect feedback mechanisms. Direct feedback mechanisms refer to intraspecific feedback, i.e., the response of plant litter or plant roots to biotic and abiotic factors in their soil and how these responses to the soil affect the performance of the plant itself or plant progeny in the community. Generally, this type of feedback is considered to be either negative (changes in the soil microbial community, physical conditions, or chemical conditions lead to a reduction in net plant growth), neutral (changes in the soil microbial community have zero net effect on the plant) or positive (changes in the soil promote the growth of the plant itself or later individuals of the same species). Indirect feedback is more commonly expressed at the community level and refers to interspecific feedback effects, i.e., the effect of feedback from one species or population on another species or population; theoretically, there is also negative, neutral, or positive feedback from indirect feedback.

On the one hand, the positive feedback is considered to be caused by plants causing changes in the soil to improve their own growth. For example, Heinze et al. (2016) found positive plant-soil feedback for fewer species while neutral plant-soil feedback for the most numerous species of three grass species in the greenhouse. On the other hand, the negative feedback is considered to be caused by plants causing changes in the soil to suppress their growth (Ehrenfeld et al. 2005). Kulmatiski and Kardol (2008) comprehensively analyzed 25 experimental plant-soil feedback papers. They found that 147 of the 271 values tested were negative feedback, with an average of 37% reduction in plant biomass. 41 of the 271 values tested were positive feedback, with an average of 43% increase in biomass. Meanwhile, the positive feedback may derive from the acquisition of nutrients and the accumulation of inter-root symbionts (Klironomos 2002, Chapman et al. 2006), while the negative feedback may be caused by nutrient depletion or the accumulation of pathogens in the soil (Berendse 1994, de Kroon et al. 2012, Van der Putten et al. 2013).

Currently, ecologists are also increasingly aware of the importance of plant-soil feedback in explaining vegetation regeneration and succession, plant community diversity and stability, and ecosystem structure and function, which have become hot research areas in community ecology and restoration ecology.

3.2. Arbuscular Mycorrhizal Fungi Affect the Direction and Intensity of Invasive Plant-Soil Feedback

The soil microbial communities potentially alter plant performance, influence population dynamics, and regulate the response of plant species to environmental change. Such a plant-soil feedback mechanism is an important aspect that affects the invasive capacity of plant species and habitat invasibility (Xi et al. 2018, Wang et al. 2019). The main mechanism of plant-soil feedback is through interactions between plants and soil microorganisms (Bever 1994, Klironomos 2002), especially AM fungi, which may be a causal driver of both positive and negative plant-soil feedback (Kivlin et al. 2018). Research suggests that plants respond differently to different AM fungi and are selective for AM fungi (Mangan et al. 2010). Therefore, changes in the community composition and diversity of AM fungi in the inter-root zone of plants can give feedback to the plants, resulting in changes in their relative growth and thus further influencing interspecific relationships between co-occurring plants (Semchenko et al. 2018, Bauer et al. 2020), and these co-occurring plants also include invasive and native plants.

AM fungi affect invasive plants in two main ways. One is to directly promote the invasive plants' growth (Wardle et al. 2002). For example, AM fungi can expand the area of an invasive plant's root system by deploying extra-rooted mycelium to create a complex network of mycelium in the soil, thus allowing it to absorb mineral nutrients from deeper and further in the soil available to the invasive plant. The other is to alter the availability of soil nutrients via the decomposition and mineralization of organic matter, thus affecting plant growth (Bardgett and Van Der Putten 2014). For instance, AM fungi can enhance soil nutrients by promoting the decomposition and nutrient release of litter (Wu et al. 2021), creating positive feedback between plants and soil (Zhao et al. 2020), which in turn promotes the invasion of exotic plants while reducing the mycorrhizal colonization rate of local plants and inhibiting the growth of local plants (Zhang et al. 2019).

Further, exotic plants can also influence AM fungi communities, and the way of plant species change the soil biota leads to positive or negative plant-soil biota feedback effects (Bever 1994, Bever 2003). For example, exotic plants influence soil biota via litter and rhizosphere inputs, and invasive plants increase AM fungal biomass by 36% via the rhizosphere pathways (Zhang et al. 2019). Moreover, exotic plants can change the soil biota in invaded areas generating positive feedback effects that facilitate invasion (Figure 2) (de Souza et al. 2019). Related native plants form positive or negative feedback (Klironomos 2002), and the direction of the feedback may influence interspecific competition and plant community composition (Hilbig and Allen 2015).

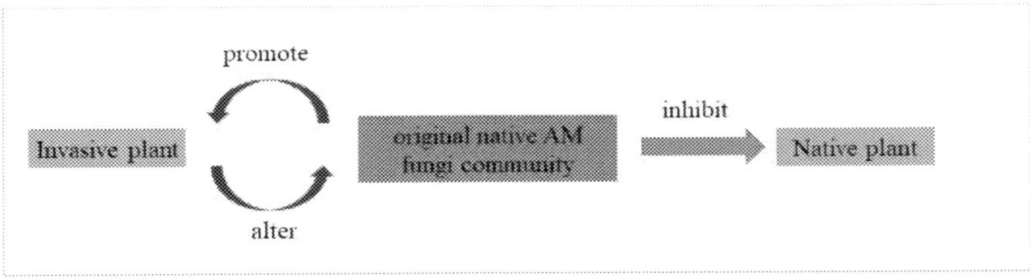

Figure 2. Framework diagram for invasive plant-soil feedback mediated by AM fungi.

AM fungi can mediate the direction and intensity of soil feedback from invasive plants (Kadowaki et al. 2018). When invasive plants compete with native plants, if the AM fungi form a stronger symbiotic relationship with the invasive plants over the native plants, they also have a higher ability to uptake water and nutrients from the soil and transport them to the invasive plants, which accelerates the colonization, growth, and reproduction of the invasive plants. Thus, the AM fungi have a direct positive feedback effect on the growth and competition of the invasive plants, promoting them to outcompete the native plants and thereby promoting the successful invasion of the alien plants (Tang et al. 2009, Harnden et al. 2011).

In addition, AM fungi do not always cause invasive plants to exhibit positive feedback in competition with native plants, and if native plants are more mycorrhizal dependent than invasive plants and form a more solid mycorrhizal symbiosis with AM fungi, the growth of native plants is promoted. Thus, in the competition between alien and local plants for growth, the AM fungi biased positive feedback towards native plants and negative feedback towards invasive plants, which inhibited plant invasion and enhanced the competitive potential of native plants (Walling and Zabinski 2006, Barto et al. 2010). In conclusion, the asymmetric distribution of benefits between the AM fungi and the host plant in the competitive growth of invasive and native plants may be the main factor behind the negative feedback or positive feedback of the AM fungi on the competitive growth of invasive plants.

3.3. The Mechanisms of Plant-Soil Feedback Regulated by Arbuscular Mycorrhizal Fungi in Invasive Competition

There are several mechanisms to interpret the different feedback effects of AM fungi on invasive plants competing with native plants. One is the mycorrhizal symbiosis hypothesis, which states that mycorrhizal fungi and host plants are interdependent, interacting, and mutually reinforcing (Liu and Wang 2003, Vogelsang and Bever 2009). Based on the strength of the mutualistic symbiosis between AM fungi and invasive plants relative to the combination with native plants, the mycorrhizal symbiosis hypothesis is further divided into the enhanced symbiosis hypothesis and the degenerative symbiosis hypothesis (Bunn et al. 2015).

On the one hand, the enhanced symbiosis hypothesis mainly reflected that AM fungi show a positive feedback effect on the growth and competition of invasive plants. Invasive plants may support more different clumped mycorrhizal types, which may provide a competitive advantage for invasive plants against native plants. Specifically, AM fungi strongly infest invasive plants and establish close symbiotic relationships superior to native plants; thus, AM fungi provide higher material and energy benefits to invasive plants than to native plants, enhancing the persistence and abundance of alien plants, leading to competitive exclusion of local plants and contributing to a successful invasion by invasive plants (Kiers and Heijden 2006, Dickie et al. 2017).

On the other hand, the degenerative symbiosis hypothesis mainly reflected that AM fungi show a negative feedback effect on the growth and competition of invasive plants. Specifically, AM fungi establish a stronger mutualistic symbiosis with local plants than invasive plants with a weak dependency on AM fungi, which increases the ability of local plants to compete against alien plants and helps local plants to resist the spreading growth of alien plants (Kisa et al. 2007). In conclusion, during competition between alien and local plants regulated by AM fungi, the direction of invasive plants' soil feedback mediated by AM fungi depends on the status of mycorrhizal symbiosis besides whether the exotic plant forms mycorrhizae with the AM fungi in the invasive soil, i.e., the mycorrhizal status of the invasive plant (Moora 2014).

The second is the novel weapons hypothesis, which states that the germination, viability, and abundance of AM fungi spores in the mycorrhizal symbiosis of native plants are inhibited by chemosensitive substances released by invasive plants through their root secretions, litter, and residue decomposition (Roberts and Anderson 2001, Meinhardt and Gehring 2012). It directly limits nutrient uptake by native plants and indirectly promotes resource acquisition by invasive plants, allowing the invasive plant a more competitive advantage (Callaway and Aschehoug 2000, Pinzone et al. 2018). The chemosynthetic disruption of invasive plant-mycorrhizal symbioses is essential in explaining the dominance of alien plants over native plants (Hale and Kalisz 2012). For instance, the alien plant *Alliaria petiolata* is gradually replacing the more mycorrhizal-dependent native plants in the invasion process (McCary et al. 2019). It is mainly due to AM fungi spores in invasive soils being affected by chemosensitizing substances (e.g., flavonoids, thioglucosides, and allyl isothiocyanate) secreted by the root of *Alliaria petiolate*, resulting in a significant reduction in the germination of local plants and in the number of mycorrhizal propagules.

The symbiotic relationship between AM fungi and local plants is disrupted, indirectly promoting the competitive growth of *Alliaria petiolate* (McCary et al. 2019). In conclusion, the growth of AM fungi is inhibited by chemosensitizing substances that exert negative feedback on native plants and positive feedback on invasive plant competition, thus affecting interspecific competition relationships. Therefore, the symbiotic relationship between native plants and AM fungi is subject to negative feedback from the chemosensory effects of invasive plants as an important mechanism determining the success of invasive plants.

Conclusion

Based on the analysis of how AM fungi regulate the growth and competition of invasive plants, this chapter expounds on the reaction of AM fungi to invasive plants competing with native plants and the feedback mechanism of mycorrhizal symbiosis and novel weapons hypothesis. Although the mycorrhizal symbionts formed by invasive plants and AM fungi play an essential role in the success of exotic plant invasion, the research on the physiological mechanism of mycorrhizal symbiosis, the nutrient utilization strategy of AM fungi on invasive plants, the feedback mechanism of nutrient competition and influencing factors is not deep enough. In the future, more work should be done on the following aspects:

1. Both greenhouse experiments and field studies need to be carried out. The current evidence for AM fungi affecting invasive plant feedbacks comes more from indoor pot experiments, where the AM fungal mycelium may be damaged during the potting process, causing it to exert a smaller feedback effect than it actually exerts in the field, and the chemosensitive material produced in indoor experiments may not adequately modulate the natural chemosensory effects experienced by neighboring plants in the field. In addition, field investigations are subject to numerous uncertain climatic influences that indoor experiments cannot control to simulate. These factors can lead to some differences in experimental results between the two experiments. Therefore, it is important to conduct indoor and field experiments to make them complementary and validate each other.
2. Further strengthen research on the physiological mechanisms of mycorrhizal symbiosis between AM fungi and host plants and the feedback mechanisms of nutrient competition between invasive plants and native plants mediated by AM fungi. The strength of mycorrhizal symbiosis and the feedback effects on nutrient uptake, distribution, and translocation directly determine the growth of invasive plants and their competition with native plants. Therefore, molecular genetic tools, isotope tracing techniques, and signal transduction knowledge should be introduced to study further the symbiotic mechanism between AM fungi and invasive plants, and to investigate the role and mechanism of AM fungi on the favorable transport and regulation of nutrients such as nitrogen and phosphorus, so as to reveal the competitive feedback mechanism of invasive plants in depth.
3. Strengthen research on the factors influencing the invasion of alien plants by AM fungi. The influence of AM fungi on the growth and competition of invasive plants is not only related to the invasive plants themselves (stage of growth and development, whether they are mycorrhizal plants, chemosensitive substances, etc.) and different

AM fungal species and sources but also influenced by external biotic factors (herbivores and other microorganisms in the soil) and abiotic factors (soil moisture, pH, nutrients, atmospheric conditions, etc.) and the interactions between these factors and AM fungi should be strengthened on The study of the feedback effects of these factors on invasive plants should be strengthened. In particular, the specific role and mechanisms of action of AM fungi on invasive alien plants in the context of global environmental change, such as increased CO_2 concentration, nitrogen and sulfur deposition, should be strengthened, and the impact on ecosystems should be studied.

4. Strengthen research on biological control techniques for invasive plants. With the increasingly serious situation of plant invasion, strengthening research on biological control of invasive plants has become the focus of ecological research. Biological competition substitution is one of the effective means of biological control. Taking advantage of the negative feedback effect of AM fungi on invasive plants, inhibiting the growth and spread of invasive plants while increasing the competitive ability of native plants, native plants can gradually replace invasive plants. In agricultural production, native plants with high economic value can be used to counteract the invasion of alien plants and improve economic efficiency. This technique can also be used for the removal of malignant invasive weeds. The selection of a suitable AM fungal inoculum is also crucial in the implementation process to control plant invasions.

Acknowledgments

This study was supported by the National Natural Science Foundation of China (NSFC: 32260268), the Science and Technology Project of Guizhou Province [(2021) General-455], and the Guizhou Hundred-level Innovative Talents Project [Qian-ke-he platform talents (2020) 6004].

References

Awaydul, A., Zhu, W., Yuan, Y., Xiao, J., Hu, H., Chen, X., Koide, R. T., and Cheng, L. (2019). Common mycorrhizal networks influence the distribution of mineral nutrients between an invasive plant, *Solidago canadensis*, and a native plant, *Kummerowa striata*. *Mycorrhiza*, 29, 29-38.

Bardgett, R. D., and Van Der Putten, W. H. (2014). Belowground biodiversity and ecosystem functioning. *Nature*, 515, 505-511.

Barto, K., Friese, C., and Cipollini, D. (2010). Arbuscular mycorrhizal fungi protect a native plant from allelopathic effects of an invader. *Journal of Chemical Ecology*, 36, 351-360.

Bauer, J. T., Koziol, L., and Bever, J. D. (2020). Local adaptation of mycorrhizae communities changes plant community composition and increases aboveground productivity. *Oecologia*, 192, 735-744.

Berendse, F. (1994). Litter decomposability--a neglected component of plant fitness. *Journal of Ecology*, 82, 187-190.

Bever, J. D. (1994). Feeback between plants and their soil communities in an old field community. *Ecology*, 75, 1965-1977.

Bever, J. D. (2003). Soil community feedback and the coexistence of competitors: conceptual frameworks and empirical tests. *New Phytologist*, 157, 465-473.

Bever, J. D., Westover, K. M., and Antonovics, J. (1997). Incorporating the soil community into plant population dynamics: the utility of the feedback approach. *Journal of Ecology*, 85, 561-573.

Bray, S. R., Kitajima, K., and Sylvia, D. M. (2003). Mycorrhizae differentially alter growth, physiology, and competitive ability of an invasive shrub. *Ecological Applications*, 13, 565-574.

Brown, B. J., Mitchell, R. J., and Graham, S. A. (2002). Competition for pollination between an invasive species (purple loosestrife) and a native congener. *Ecology*, 83, 2328-2336.

Bunn, R. A., Ramsey, P. W., and Lekberg, Y. (2015). Do native and invasive plants differ in their interactions with arbuscular mycorrhizal fungi? A meta-analysis. *Journal of Ecology*, 103, 1547-1556.

Callaway, R. M., and Aschehoug, E. T. (2000). Invasive plants versus their new and old neighbors: a mechanism for exotic invasion. *Science*, 290, 521-523.

Chapman, S. K., Langley, J. A., Hart, S. C., and Koch, G. W. (2006). Plants actively control nitrogen cycling: uncorking the microbial bottleneck. *New Phytologist*, 169, 27-34.

Cheng, J. K., Yue, M. F., Yang, H. R., Chen, B. M., and Xin, G. R. (2019). Do arbuscular mycorrhizal fungi help the native species *Bidens biternata* resist the invasion of *Bidens alba*? *Plant and Soil*, 444, 443-455.

Cuda, J., Skalova, H., Janovsky, Z., and Pysek, P. (2015). Competition among native and invasive Impatiens species: the roles of environmental factors, population density and life stage. *Aob Plants*, 7.

de Kroon, H., Hendriks, M., van Ruijven, J., Ravenek, J., Padilla, F. M., Jongejans, E., Visser, E. J., and Mommer, L. (2012). Root responses to nutrients and soil biota: drivers of species coexistence and ecosystem productivity. *Journal of Ecology*, 100, 6-15.

de Souza, T. A. F., Santos, D., de Andrade, L. A., and Freitas, H. (2019). Plant-soil feedback of two legume species in semi-arid Brazil. *Brazilian Journal of Microbiology*, 50, 1011-1020.

Diagne, N., Ngom, M., Djighaly, P. I., Fall, D., Hocher, V., and Svistoonoff, S. (2020). Roles of arbuscular mycorrhizal fungi on plant growth and performance: Importance in biotic and abiotic stressed regulation. *Diversity*, 12, 370.

Dickie, I. A., Bufford, J. L., Cobb, R. C., Desprez-Loustau, M. L., Grelet, G., Hulme, P. E., Klironomos, J., Makiola, A., Nuñez, M. A., and Pringle, A. (2017). The emerging science of linked plant–fungal invasions. *New Phytologist*, 215, 1314-1332.

Dukes, J. S., and Mooney, H. A. (1999). Does global change increase the success of biological invaders? *Trends in Ecology and Evolution*, 14, 135-139.

Ehrenfeld, J. G. (2010). Ecosystem consequences of biological invasions. *Annual Review of Ecology, Evolution, and Systematics*, 41, 59-80.

Ehrenfeld, J. G., Ravit, B., and Elgersma, K. (2005). Feedback in the plant-soil system. *Annual Review of Environment and Resources*, 30, 75-115.

Eppinga, M. B., Rietkerk, M., Dekker, S. C., De Ruiter, P. C., Van der Putten, W. H., and Van der Putten, W. H. (2006). Accumulation of local pathogens: a new hypothesis to explain exotic plant invasions. *Oikos*, 114, 168-176.

Fellbaum, C. R., Gachomo, E. W., Beesetty, Y., Choudhari, S., Strahan, G. D., Pfeffer, P. E., Kiers, E. T., and Bücking, H. (2012). Carbon availability triggers fungal nitrogen uptake and transport in arbuscular mycorrhizal symbiosis. *Proceedings of the National Academy of Sciences*, 109, 2666-2671.

Fellbaum, C. R., Mensah, J. A., Cloos, A. J., Strahan, G. E., Pfeffer, P. E., Kiers, E. T., and Bücking, H. (2014). Fungal nutrient allocation in common mycorrhizal networks is regulated by the carbon source strength of individual host plants. *New Phytologist*, 203, 646-656.

Guo, Y., He, Y., Wu, B., Lin, Y., He, M., Han, X., Xia, T., Shen, K., Kang, L., and Tan, Q. (2021). The interspecific competition presents greater nutrient facilitation compared to intraspecific competition through AM fungi interacting with litter for two host plants in karst soil. *Journal of Plant Ecology*, 15, 399-412.

Hale, A. N., and Kalisz, S. (2012). Perspectives on allelopathic disruption of plant mutualisms: a framework for individual-and population-level fitness consequences. *Plant Ecology*, 213, 1991-2006.

Harnden, J., MacDougall, A. S., and Sikes, B. A. (2011). Field-based effects of allelopathy in invaded tallgrass prairie. *Botany*, 89, 227-234.

He, Y. (2019). *Ecological Adaptability of Karst Plants Regulated by Arbuscular Mycorrhizae* (In Chinese). Science Press, Beijing.

He, Y. J., Cornelissen, J. H., Wang, P. P., Dong, M., and Ou, J. (2019). Nitrogen transfer from one plant to another depends on plant biomass production between conspecific and heterospecific species via a common arbuscular mycorrhizal network. *Environmental Science and Pollution Research*, 26, 8828-8837.

He, Y. J., and Zhong, Z. C. (2011). Drought Resistance of *Cinnamomum camphora* Seedlings Inoculated with Different AM Fungi in Karst Soil. *Bulletin of Botanical Research*, 31, 597-530.

He, Y., Liu, J. C., and Liu, J. M. (2008). Photosynthetic characteristics of *Broussonetia papyrifera* seedlings inoculated AM fungus in limestone soil substratum. *Bulletin of Botanical Research*, 28, 452-457.

Heinze, J., Sitte, M., Schindhelm, A., Wright, J., and Joshi, J. (2016). Plant-soil feedbacks: a comparative study on the relative importance of soil feedbacks in the greenhouse versus the field. *Oecologia*, 181, 559-569.

Hilbig, B. E., and Allen, E. B. (2015). Plant-soil feedbacks and competitive interactions between invasive Bromus diandrus and native forb species. *Plant and Soil*, 392, 191-203.

Hodge, A., and Fitter, A. H. (2010). Substantial nitrogen acquisition by arbuscular mycorrhizal fungi from organic material has implications for N cycling. *Proceedings of the National Academy of Sciences*, 107, 13754-13759.

Huang, K., Kong, D. L., Lu, X. R., Feng, W. W., Liu, M. C., and Feng, Y. L. (2020). Lesser leaf herbivore damage and structural defense and greater nutrient concentrations for invasive alien plants: Evidence from 47 pairs of invasive and non-invasive plants. *Science of the Total Environment*, 723, 137829.

Johnson, N. C., Wilson, G. W., Wilson, J. A., Miller, R. M., and Bowker, M. A. (2015). Mycorrhizal phenotypes and the L aw of the Minimum. *New Phytologist*, 205, 1473-1484.

Kadowaki, K., Yamamoto, S., Sato, H., Tanabe, A. S., Hidaka, A., and Toju, H. (2018). Mycorrhizal fungi mediate the direction and strength of plant–soil feedbacks differently between arbuscular mycorrhizal and ectomycorrhizal communities. *Communications Biology*, 1, 1-11.

Kiers, E. T., Duhamel, M., Beesetty, Y., Mensah, J. A., Franken, O., Verbruggen, E., Fellbaum, C. R., Kowalchuk, G. A., Hart, M. M., and Bago, A. (2011). Reciprocal rewards stabilize cooperation in the mycorrhizal symbiosis. *Science*, 333, 880-882.

Kiers, E. T., and Heijden, M. G. v. d. (2006). Mutualistic stability in the arbuscular mycorrhizal symbiosis: exploring hypotheses of evolutionary cooperation. *Ecology*, 87, 1627-1636.

Kisa, M., Sanon, A., Thioulouse, J., Assigbetse, K., Sylla, S., Spichiger, R., Dieng, L., Berthelin, J., Prin, Y., and Galiana, A. (2007). Arbuscular mycorrhizal symbiosis can counterbalance the negative influence of the exotic tree species *Eucalyptus camaldulensis* on the structure and functioning of soil microbial communities in a sahelian soil. *FEMS Microbiology Ecology*, 62, 32-44.

Kivlin, S. N., Bedoya, R., and Hawkes, C. V. (2018). Heterogeneity in arbuscular mycorrhizal fungal communities may contribute to inconsistent plant-soil feedback in a Neotropical forest. *Plant and Soil*, 432, 29-44.

Klironomos, J. N. (2002). Feedback with soil biota contributes to plant rarity and invasiveness in communities. *Nature*, 417, 67-70.

Kroon, H. D., Hendriks, M., Ruijven, J. V., Ravenek, J., Padilla, F. M., Jongejans, E., Visser, E. J. W., and Mommer, L. (2015). Root responses to nutrients and soil biota: drivers of species coexistence and ecosystem productivity. *Journal of Ecology*, 100, 6-15.

Kulmatiski, A., and Kardol, P. (2008). Getting plant—soil feedbacks out of the greenhouse: experimental and conceptual approaches. Pages 449-472, in *Progress in Botany*. Springer.

Liu, R., and Wang, F. (2003). Selection of appropriate host plants used in trap culture of arbuscular mycorrhizal fungi. *Mycorrhiza*, 13, 123-127.

Lodge, D. M. (1993). Biological invasions: lessons for ecology. *Trends in Ecology and Evolution*, 8, 133-137.

Mack, K. M., and Bever, J. D. (2014). Coexistence and relative abundance in plant communities are determined by feedbacks when the scale of feedback and dispersal is local. *Journal of Ecology*, 102, 1195-1201.

Mangan, S. A., Herre, E. A., and Bever, J. D. (2010). Specificity between Neotropical tree seedlings and their fungal mutualists leads to plant–soil feedback. *Ecology*, 91, 2594-2603.

Mangla, S., Inderjit, and Callaway, R. M. (2008). Exotic invasive plant accumulates native soil pathogens which inhibit native plants. *Journal of Ecology*, 96, 58-67.

McCary, M. A., Zellner, M., and Wise, D. H. (2019). The role of plant–mycorrhizal mutualisms in deterring plant invasions: Insights from an individual-based model. *Ecology and Evolution*, 9, 2018-2030.

Meinhardt, K. A., and Gehring, C. A. (2012). Disrupting mycorrhizal mutualisms: a potential mechanism by which exotic tamarisk outcompetes native cottonwoods. *Ecological Applications*, 22, 532-549.

Merrild, M. P., Ambus, P., Rosendahl, S., and Jakobsen, I. (2013). Common arbuscular mycorrhizal networks amplify competition for phosphorus between seedlings and established plants. *New Phytologist*, 200, 229-240.

Molina, R., and Horton, T. R. (2015). Mycorrhiza specificity: its role in the development and function of common mycelial networks. Pages 1-39, in *Mycorrhizal Networks*. Springer.

Moora, M. (2014). Mycorrhizal traits and plant communities: perspectives for integration. *Journal of Vegetation Science*, 25, 1126-1132.

Neuenkamp, L., Zobel, M., Lind, E., Gerz, M., and Moora, M. (2019). Arbuscular mycorrhizal fungal community composition determines the competitive response of two grassland forbs. *PLoS One*, 14, e0219527.

Pejchar, L., and Mooney, H. A. (2009). Invasive species, ecosystem services and human well-being. *Trends in Ecology and Evolution*, 24, 497-504.

Pinzone, P., Potts, D., Pettibone, G., and Warren, R. (2018). Do novel weapons that degrade mycorrhizal mutualisms promote species invasion? *Plant Ecology*, 219, 539-548.

Rahman, M. K. U., Zhou, X. G., and Wu, F. Z. (2019). The role of root exudates, CMNs, and VOCs in plant-plant interaction. *Journal of Plant Interactions*, 14, 630-636.

Řezáčová, V., Řezáč, M., Gryndlerová, H., Wilson, G. W., and Michalová, T. (2020). Arbuscular mycorrhizal fungi favor invasive *Echinops sphaerocephalus* when grown in competition with native *Inula conyzae*. *Scientific Reports*, 10, 1-10.

Roberts, K. J., and Anderson, R. C. (2001). Effect of garlic mustard [*Alliaria petiolata* (Beib. Cavara & Grande)] extracts on plants and arbuscular mycorrhizal (AM) fungi. *The American Midland Naturalist*, 146, 146-152.

Scheublin, T. R., Van Logtestijn, R. S., and Van Der Heijden, M. G. (2007). Presence and identity of arbuscular mycorrhizal fungi influence competitive interactions between plant species. *Journal of Ecology*, 95, 631-638.

Schott, S., Valdebenito, B., Bustos, D., Gomez-Porras, J. L., Sharma, T., and Dreyer, I. (2016). Cooperation through competition - dynamics and microeconomics of a minimal nutrient trade system in arbuscular mycorrhizal symbiosis. *Frontiers in Plant Science*, 7, 912.

Semchenko, M., Leff, J. W., Lozano, Y. M., Saar, S., Davison, J., Wilkinson, A., Jackson, B. G., Pritchard, W. J., De Long, J. R., and Oakley, S. (2018). Fungal diversity regulates plant-soil feedbacks in temperate grassland. *Science Advances*, 4, eaau4578.

Shen, K. P., Cornelissen, J. H. C., Wang, Y. J., Wu, C. B., He, Y. J., Ou, J., Tan, Q. Y., Xia, T. T., Kang, L. L., Guo, Y., and Wu, B. L. (2020). AM fungi alleviate phosphorus limitation and enhance nutrient competitiveness of invasive plants via mycorrhizal networks in Karst areas. *Frontiers in Ecology and Evolution*, 8, 125.

Simard, S. W., Jones, M. D., and Durall, D. M. (2003). Carbon and nutrient fluxes within and between mycorrhizal plants. Pages 33-74, in *Mycorrhizal Ecology*. Springer.

Simberloff, D., Martin, J. L., Genovesi, P., Maris, V., Wardle, D. A., Aronson, J., Courchamp, F., Galil, B., García-Berthou, E., and Pascal, M. (2013). Impacts of biological invasions: what's what and the way forward. *Trends in Ecology and Evolution*, 28, 58-66.

Smith, S. E., and F. A. (2011). Roles of arbuscular mycorrhizas in plant nutrition and growth: New paradigms from cellular to ecosystem scales. *Annual Review Plant Biology*, 2011, 62, 227-250.

Smith, S. E., and Read, D. J. (2008). Mycorrhizal Symbiosis. *Quarterly Review of Biology*, 3, 273-281.

Stock, S. C., Koester, M., Boy, J., Godoy, R., Nájera, F., Matus, F., Merino, C., Abdallah, K., Leuschner, C., and Spielvogel, S. (2021). Plant carbon investment in fine roots and arbuscular mycorrhizal fungi: A cross-biome study on nutrient acquisition strategies. *Science of the Total Environment*, 781, 146748.

Szymura, M., and Szymura, T. H. (2016). Interactions between alien goldenrods (Solidago and Euthamia species) and comparison with native species in Central Europe. *Flora*, 218, 51-61.

Tamme, R., Hiiesalu, I., Laanisto, L., Szava-Kovats, R., and rtel, M. P. (2010). Environmental heterogeneity, species diversity and co-existence at different spatial scales. *Journal of Vegetation Science*, 21, 796-801.

Tang, J., Zhang, Q., Yang, R., and Chen, X. (2009). Effects of exotic plant *Solidago canadensis* L. on local arbuscular mycorrhizal fungi. *Bull Sci Tech*, 25, 233-237.

Theoharides, K. A., and Dukes, J. S. (2007). Plant invasion across space and time: factors affecting nonindigenous species success during four stages of invasion. *New Phytologist*, 176, 256-273.

Van der Putten, W. H., Bardgett, R. D., Bever, J. D., Bezemer, T. M., Casper, B. B., Fukami, T., Kardol, P., Klironomos, J. N., Kulmatiski, A., and Schweitzer, J. A. (2013). Plant–soil feedbacks: the past, the present and future challenges. *Journal of Ecology*, 101, 265-276.

van Kleunen, M., Dawson, W., Essl, F., Pergl, J., Winter, M., Weber, E., Kreft, H., Weigelt, P., Kartesz, J., Nishino, M., Antonova, L. A., Barcelona, J. F., Cabezas, F. J., Cardenas, D., Cardenas-Toro, J., Castano, N., Chacon, E., Chatelain, C., Ebel, A. L., Figueiredo, E., Fuentes, N., Groom, Q. J., Henderson, L., Inderjit, Kupriyanov, A., Masciadri, S., Meerman, J., Morozova, O., Moser, D., Nickrent, D. L., Patzelt, A., Pelser, P. B., Baptiste, M. P., Poopath, M., Schulze, M., Seebens, H., Shu, W. S.,

Thomas, J., Velayos, M., Wieringa, J. J., and Pysek, P. (2015). Global exchange and accumulation of non-native plants. *Nature*, 525, 100-103.

Vogelsang, K. M., and Bever, J. D. (2009). Mycorrhizal densities decline in association with nonnative plants and contribute to plant invasion. *Ecology*, 90, 399-407.

Walling, S. Z., and Zabinski, C. A. (2006). Defoliation effects on arbuscular mycorrhizae and plant growth of two native bunchgrasses and an invasive forb. *Applied Soil Ecology*, 32, 111-117.

Wang, M., Ruan, W., Kostenko, O., Carvalho, S., Hannula, S. E., Mulder, P. P. J., Bu, F., van der Putten, W. H., and Bezemer, T. M. (2019). Removal of soil biota alters soil feedback effects on plant growth and defense chemistry. *New Phytologist*, 221, 1478-1491.

Wardle, D. A., Bardgett, R. D., Klironomos, J. N., Setala, H., and Van, W. H. (2004). Ecological linkages between aboveground and belowground biota. *Science*, 304, 1629-1633.

Wardle, D., Bonner, K., and Barker, G. (2002). Linkages between plant litter decomposition, litter quality, and vegetation responses to herbivores. *Functional Ecology*, 16, 585-595.

Weidenhamer, J. D., and Callaway, R. M. (2010). Direct and indirect effects of invasive plants on soil chemistry and ecosystem function. *Journal of Chemical Ecology*, 36, 59-69.

Wilson, G. W. T., Hickman, K. R., and Williamson, M. M. (2012). Invasive warm-season grasses reduce mycorrhizal root colonization and biomass production of native prairie grasses. *Mycorrhiza*, 22, 327-336.

Workman, R. E. (2014). *Effects of Arbuscular Mycorrhizal Fungal Infection and Common Mycelial Network Formation on Invasive Plant Competition*. Portland State University.

Workman, R. E., and Cruzan, M. B. (2016). Common mycelial networks impact competition in an invasive grass. *American Journal of Botany*, 103, 1041-1049.

Wu, B., Guo, Y., He, M., Han, X., Zang, L., Liu, Q., Chen, D., Xia, T., Shen, K., Kang, L., and He, Y. (2021). AM fungi endow greater plant biomass and soil nutrients under interspecific competition rather than nutrient releases for litter., *Forests*, 12, 1704.

Wu, Q., Tang, Y., Dong, T., Liao, Y., Li, D., He, X., and Xu, X. (2018). Additional AM fungi inoculation increase *Populus cathayana Intersexual* competition. *Frontiers in Plant Science*, 9, 607.

Xi, N., Chu, C., and Bloor, J. M. G. (2018). Plant drought resistance is mediated by soil microbial community structure and soil-plant feedbacks in a savanna tree species. *Environmental and Experimental Botany*, 155, 695-701.

Xia, T., Wang, Y., He, Y., Wu, C., Shen, K., Tan, Q., Kang, L., Guo, Y., Wu, B., and Han, X. (2020). An invasive plant experiences greater benefits of root morphology from enhancing nutrient competition associated with arbuscular mycorrhizae in karst soil than a native plant. *PLoS One*, 15, e0234410.

Yang, G., Liu, N., Lu, W., Wang, S., Kan, H., Zhang, Y., Xu, L., and Chen, Y. (2014). The interaction between arbuscular mycorrhizal fungi and soil phosphorus availability influences plant community productivity and ecosystem stability. *Journal of Ecology*, 102, 1072-1082.

Yu, H., and He, W. (2021). Arbuscular Mycorrhizal Fungi Compete Asymmetrically for *Amino Acids* with Native and Invasive *Solidago. Microbial Ecology*, 1-10.

Zhang, F. J., Li, Q., Chen, F. X., Xu, H. Y., and Wan, F. H. (2017). Arbuscular mycorrhizal fungi facilitate growth and competitive ability of an exotic species *Flaveria bidentis*. *Soil Biology and Biochemistry*, 115, 275-284.

Zhang, P., Li, B., Wu, J., and Hu, S. (2019). Invasive plants differentially affect soil biota through litter and rhizosphere pathways: a meta-analysis. *Ecology Letters*, 22, 200-210.

Zhao, Y. Z., Liu, M. C., Feng, Y. L., Wang, D., Feng, W. W., Clay, K., Durden, L. A., Lu, X. R., Wang, S., and Wei, X. L. (2020). Release from below-and aboveground natural enemies contributes to invasion success of a temperate invader. *Plant and Soil*, 452, 19-28.

Chapter 4

The Strategic Combination of Mycorrhizal Inoculants, Fertilizers and Green Manures Improve Crop Productivity. Review of Cuban Research

Ramón Rivera Espinosa[1,*]
Pedro José González Cañizares[1]
Luis Ruiz Martínez[2]
Gloria Marta Martin Alonso[1]
and Adriano Cabrera Rodríguez[1]
[1] National Institute of Agricultural Science, San Jose de las Lajas, Cuba
[2] Research Institute on Tropical Root Crops, Villa Clara, Cuba

Abstract

From the different ecoservices in which the arbuscular mycorrhizal symbiosis participates, the conscious management of this symbiosis has a high potential for agriculture. Inoculation can be one of the ways to achieve its benefits, but it is necessary to find the factors on which its effectiveness depends, integrate them with agricultural practices and validate them on a productive scale.

A summary of the work carried out with the objectives of integrating mycorrhizal inoculant management into the nutrient supply schemes and technologies of different crops on various types of soils is presented. The inoculants were applied via seed coatings. The effectiveness of three "simple" inoculants formulated from generalist isolates with the plant

* Corresponding Author's Email: rrabracadabra14@gmail.com.

In: New Research on Mycorrhizal Fungus
Editors: Qiang-Sheng Wu, Ying-Ning Zou, Yue-Jun He et al.
ISBN: 979-8-88697-637-3

species was found to be dependent on soil conditions and to a high extent on soil pH, and ranges of soil pH were established in which they were effective. Inoculation allowed high yields, mycorrhizal functioning and satisfactory macronutrient concentrations in the crops, with lower fertilizer requirements and better fertilizer utilisation. The effectiveness of the inoculation depended on the nutrient supply and requires a sub-optimal supply, although it can be of only one of the macronutrients. Inoculation of green manures increases the potentiality of green manures, further decreases the amounts of fertilizers for the main crops and becomes a way to effectively mycorrhizate successional or associated crops, with high practical importance in vegetatively propagated crops or in established plantations.

The management of inoculated systems requires monitoring through soil analysis and balances of nutrient inputs and exports, which allows the supply of nutrients in successive harvests to be adjusted. Validation actions of this integrated management with farmers in fodder production was satisfactory, corroborating the experimental information obtained and endorsing their use in agricultural practice.

Keywords: effectiveness of inoculants, crops, soils, integrated plant nutrition systems, canavalia

1. Introduction

The rational supply of nutrients in the agroecosystems is a necessity to guarantee adequate yields and maintenance of soil fertility (Van Lauwe et al., 2010) as well as avoiding environmental pollution associated with the indiscriminate use of fertilizers (Tilman et al., 2002).Although in recent years has become clear the importance of establishing technologies that enhance the biological activity and the use of microorganisms and that allow to use smaller amounts of inputs (Arora et al., 2016; Le Mire et al., 2016; Schütz et al., 2018), its use is still limited and fundamentally restrained to the applications of rhizobia (Lesueur et al., 2016). For years, high fertilizer prices have a negative impact on a global scale and in Cuba and many developing countries low percentages of agricultural areas are fertilized, which causes negative effects on yields and soil fertility. Both causes have driven R & D & I aimed at the use of bioproducts and organic-mineral sources in sustainable agricultural production technologies (The Royal Society, 2009). This situation has become more acute since the COVID19 pandemic.

There is no doubt in the universality of the mycorrhizal symbiosis and the benefits they bring to plants and the environment associated with the increase in nutrient and water intake (Ortas and Rafique, 2017; Ruiz-Lozano et al., 2012), the effect of bioprotection against some radical and foliar pests (Pozo et al., 2009), tolerance to heavy metals, as well as other different ecoservices among which are increases in soil aggregates (Lehman et al., 2017), cooperation with others microorganisms rizospheric and participation in the C and N cycles (Hodge and Storer, 2015; Jansa et al., 2019; Parihar et al., 2020) and improvements in agroecosystem resilience being a current challenge the management of this symbiosis to optimize its benefits in the agroecosystems. (Rillig et al., 2019).

Although this symbiosis has the characteristic that a defining part of its structure is found in the soil, so the importance of the edaphic environment for the functioning of mycorrhizae is not surprising (Helgason and Fitter, 2009) however for many years a phytocentric approach has prevailed (Hamel and Strullu, 2006). In recent years a growing recognition of the importance of edaphic conditions on the distribution and frequency of resident arbuscular mycorrhizal fungal (AMF) species in agroecosystems has begun (Hamel et al., 1997; Oehl et al., 2010; Santos-González et al., 2011; Sikes et al., 2014), however there is little information on the influence of soil properties on the effectiveness of inoculants. In principle, there are two ways to enhance arbuscular mycorrhizal functioning in the crops, either by managing resident mycorrhization or by inoculating efficient strains of arbuscular mycorrhizal fungi (AMF).

Although positive reports on the inoculation of AMF isolates have increased in recent years and several published metadata make it clear that they are highly effective (Hoeksema et al., 2010; Pellegrino et al., 2015; Schütz et al., 2018; Zhang et al., 2018), there are few publications that allow comprehensive recommendations on management of AMF inoculants, ways to apply them and integration with cultural practices and fertilizers in agricultural production technologies.

Rodríguez and Sanders (2015) have pointed out that in non-sterile conditions the efficiency of an AMF isolate cannot be established since the positive response to its application may be due either to the inoculated isolate (effect direct) or to the interaction with the resident mycorrhiza or both. On the other hand, the integration of mycorrhizal inoculants in agricultural production technologies, with the exception of some systems for the production of seedlings or acclimatization of vitroplants, should not require sterilizing the soil, whose practice would be economically and environmentally prohibitive.

Even with the uncertainty between a "direct or indirect" beneficial effect, the integration of mycorrhizal inoculants in crops technologies requires determining under non-sterilization conditions whether or not there are regularities with the application of the different AMF isolates (inoculants) and the factors that influence in these. These regularities will allow the establishment of criteria for recommending which inoculants to apply in cultures dependent on mycorrhization.

The main objective of the work program was to integrate mycorrhizal inoculants into the nutrient supply systems and crop technologies and consequently achieve reductions in the amounts of fertilizers. At the beginning of the programme at the end of the last century, few results were published in this sense, mainly in the production of coffee or citrus seedlings and in the cultivation of cassava (Kang et al., 1980, López et al., 1984; Sieverding 1991; Siqueira et al., 1993) and even today there are still few field studies carried out for these purposes.

Among the causes of the low number of results was the non-existence of inoculants that are applied in low quantities (< 10 kg ha^{-1}) with reproducible results. The development of a "simple" inoculant in solid formulation (Fernandez et al., 2000) that is applied in quantities equivalent to 6 to 10% of the seed, via seed coatings (Rivera and Fernandez, 2006), which can also be applied by coating the vegetative propagules (Ruiz et al., 2012) and directly under the seed in coffee plant seedlings and acclimatisation of vitroplantas (Rivera et al., 2003; Simó et al., 2017), made it possible promoting experiments, with outputs applicable on a productive scale, in different crops and soil types with the objectives of integrating mycorrhizal inoculants with nutrient supply schemes.

The success of mycorrhizal inoculants requires their integration with different cultural practices. Green manures present a whole series of benefits associated with nitrogen inputs, nutrient recycling, soil cover and erosion reduction among others (Garcia et al., 2000 and 2002, Crews and People, 2004; Fernandes de Sousa et al., 2016; Vargas et al., 2017). As mycorrhizal crops increase their growth not only because of the already explained benefits of mycorrhizae, but also because the mycorrhizal function itself in leguminous plants increases the effectiveness of biological nitrogen fixation (Yasmen et al., 2012; Larimer et al., 2014)

Moreover, from positive results on the permanence effect of the inoculation made to a preceding crop on the successor crop mycorrhization

(Rivera and Fernández, 2003), it was considered important to study the inclusion of inoculated green manures in the development of integrated plant nutrition systems, not only for its contribution in the nutrition of the subsequent crop, but because it could be an effective way for the mycorrhization of this one.

The present chapter is a summary of the main results obtained aimed at demonstrating the feasibility and benefits of the management and integration of mycorrhizal inoculants in nutrient supply schemes and crop technologies, with emphasis in this case on the integration with green manures and fertilizers and in which different components of this management are discussed in a stepwise manner and giving answers to the following questions:

1) Is there any regularity in crop response to the inoculation of different AMF isolates and what does this regularity depend on?
2) How does the application of effective inoculants modify the amounts of fertilizers to be applied to crops to obtain high yields?
3) Does co-management of green manures and mycorrhizal inoculants enhance fertilizer efficiency and mycorrhizal performance of the main crop?

2. Cuban Soils

The Cuban territory is an archipelago located in the Caribbean Sea, at the entrance of the Gulf of Mexico between 74° and 84° west longitude and between 19° and 23° north latitude, with a seasonally humid tropical warm climate with maritime influence. The average temperature ranges from 24° Celsius in the plains to 26° Celsius and above on the eastern coasts, decreasing in the few and not large mountain systems. The average annual accumulated rainfall is 1335 mm divided into a rainy period from May to October, where approximately 80% of the annual total is registered, and the little rainy period from November to April, with the remaining 20%.

There is a diversity of soil-forming rocks in the country, such as serpentinites, hard limestone and acid igneous rocks, which generally do not exceed 200 million years old (Iturralde et al., 2012). This rock diversity has made the archipelago's soil cover very heterogeneous, occupying low coastal plains and high plains (slightly more than 75% of the territory), undulating regions of the piedmont and hilly premontane regions. Such cover is composed of soils ranging from highly evolved soils such as Ferrasol, Nitisol, Lixisol according to WRB, (2014), to soils in which minerals are inherited and there is usually no soil disturbance. Among the Soil Groupings, there are marked differences in terms of their properties; even more, within the same Grouping, types and subtypes differ substantially (Cabrera, 1991, Muñoz, 1997).

Nevertheless, soils formed from calcareous materials of marine origin predominate, in which more than pH-H_2O exceeds 6 in 80%; about 65% of the agricultural soils have an exchangeable Ca concentration above 10 $cmol_c\ kg^{-1}$ and in some groupings it exceeds 40 $cmol_c\ kg^{-1}$; while about 50% of soils have an exchangeable Mg concentration above 5 $cmol_c\ kg^{-1}$. Even relatively evolved soils such as several of Nitisols types possess an exchange capacity up to 20 $cmol_c\ kg^{-1}$ and between 10 to 12 $cmol_c\ kg^{-1}$ of exchangeable Ca (Hernández-Jiménez et al., 2015).

There are three properties that stand out and among all, mark a substantial difference with other soils from tropical regions: only about 6% possess exchangeable acidity above 1 $cmol_c\ kg^{-1}$, about 4% possess an exchangeable Al concentration higher than 1 $cmol_c\ kg^{-1}$ and the average maximum P adsorption capacity is 246 $\mu g\ P\ g^{-1}$ of soil. These properties present lower values than those found in other areas of the tropics and contribute to the fact that Cuban soils potentially possess greater fertility. According to the criteria expressed by Lambers et al. (2008), they could be grouped in a zone of appropriate edaphic conditions for the functioning of arbuscular mycorrhizae.

Table 1 shows the main properties of different types of agricultural soils and their management in which the different experiments were carried out. In several of the soils and especially in the available P content, the influence of P fertilisation can be observed. Also, contents of resident mycorrhizal spores are shown, which are generally low and possibly associated with the high levels of fertilization in Cuban agriculture in the period 1960-1990 and the negative effects of these and different soil degradation processes on the resident mycorrhizal population (Hamel and Strullu, 2006; Guo et al., 2020).

However, in studies carried out in different ecosystems throughout the country, a wide diversity of AMF has been found. In this sense Furralzola et al. (2021) summarises the different works carried out:

> The species reported in Cuba belong to 11 families and 25 genera. Glomeraceae (37%) and Acaulosporaceae (19%) are the predominant families and the genera *Glomus* (18%) and *Acaulospora* (18%). The species most commonly reported in Cuba, independent of the ecosystem, *are Acaulospora scrobiculata, Diversispora spurca, Funneliformis mosseae, Glomus pachycaule, Rhizoglomus intraradices and Sclerocystis sinuosa*. These could be considered generalist species adapted to different environments.

Table 1. Types of soil, location, province and main initial properties of soils in the experiments realized during 1995-2018

Soils (WRB, 2014) and location	Land use	pH	MO	P	K	Ca	Mg	Na	Resident AMF spores in 50 g of soil
		H_2O	(g kg^{-1})	(mg kg^{-1})	($cmol_c$ kg^{-1})				
Vertic Phaozem/22° 35' N 80°18' O/Villa Clara	Miscellaneous crops fertilized	7.8 – 7.9	20.8	3.14	0.49	44.6	3.21	n.d	50
Eutric Cambisol/22° 35' N 80°18' O/Villa Clara		7.3- 7.6	21.6	32	0.60	44.8	3.2	n.d	47
Gley Pelic Vertisol/22° 55' 59" N 82° 22' 39" O/Artemisa	Fertilized forages	7.3	41.3	5.6	0.69	37.2	3.5-7	0.55	312
Eutric Cambisol/20° 09' 04.7" N 74° 49'23.5" O/Guantánamo	Miscellaneous crops fertilized	7.3	28-30	21.6	0.6	40.0	8.0	n.d	n.d
Lixic Eutric Nitisol/23° 01'N82° 08'O Mayabeque		6.50-6.95	32.5	8-100	0.21	9.7-11	2.2	0.15	109
Lixic Eutric Nitisol/22° 55' 59" N 82° 22' 39" O/Artemisa	Fertilized forages	6.3	32.5	11	0.34	9.7	2.2	0.15	100
Rhodic Alumic Acrisol/22° 16' 55" N 83° 49'19" O/Pinar del Rio	Fertilized tobacco	6.00	13.6-22	100	0.44	4.7	1.9	0.10	50
Eutric Cambisol/20°09′ N 76°16′ O/Santiago de Cuba	Miscellaneous crops	6.0-6.2	26.1	27	0.57	27.0	4.0	n.d	n.d
Dystric Cambisol/22° 09′ N 80° 12′ O/Cienfuegos	Forage	4.9-5.3	17.8	0.9	0.2	6	2.2	0.23	80
Albic Pisoplintic Plintosol/22° 57' N 83° 23' O/Pinar del Rio	Rice flooded soil	7.2	28.6	42	0.18	6.97	3.1	0.2	n.d
Albic Pisoplintic Plintosol/22° 39' 33" N80° 29' 46" O/VillaClara	Limed forage	6.0-7.2	26.0	7.4	0.12	4.2-6.0	1.3	0.11	69
Albic Pisoplintic Plintosol/22° 39' 33" N80° 29' 46" O/VillaClara	Forage	4.7 – 4.9	24.3	6.8	0.11	3.5	1.2	0.10	69

n.d.: not determined. pH in H_2O in soil:solution relation (1:2.5). Determination of organic matter level by the Walkley-Black method. Available P with extractive solution of$H_2SO_4$0.025 M, and color developing with ammonium molibdate. Extraction of exchangeable cations with NH_4Ac 1 M and pH 7 in soil: solution relation of 1:5. In all the casesaccording to the "Handbook of Analytical Techniques for soil and foliar analyses, organic manure and chemical fertilizers" (Paneque et al., 2010). Spores by Gedermann and Nicholson (1963).

3. Effectiveness of AMF Inoculants and Recommendation Criteria

In the last decade of the last century, several studies were carried out in Cuba to evaluate the effectiveness of different AMF isolates when inoculated on various crops and soil types under non-sterile conditions. The AMF isolates were always applied as single inoculants. Three main assumptions were made: inoculation as a way to achieve effective mycorrhization of crops, the soil type importance in the effectiveness of inoculated isolates and the nutrient supply influence on the symbiosis effectiveness.

The first group of these experiments were carried out in the production of coffee plant seedlings (Fernandez, 1999; Sánchez et al., 2000; Sánchez et al., 2006 a, b, c) in which between 6 and 14 AMF isolates inoculated in six types of soils and in combination with different levels of nutrient supply, given by various soil: organic fertilizer ratios in the substrates (3/1, 5/1 and 7/1) were compared. All experiments were repeated three times. In all these experiments there was a positive and reproducible response to inoculation, with differences between the effectiveness of inoculated isolates, so that isolates that originated the greatest response were different for each type of soil. The inoculated isolate that produced the highest responses expressed in leaf area and colonization were referred to as efficient isolates. However, for the same type of soil there may be more than one efficient isolate.

In addition, the response to inoculation of the efficient isolates also depended on the soil/organic fertilizer ratio of the substrate. The greatest response was found in almost all soils with soil/organic fertilizer ratios of 5/1 (Rivera et al., 2003), which contain lower amounts of organic fertilizers than the 3:1 ratio used to obtain coffee plant seedlings in Cuba (MINAG, 1987). Also, to the extent that soil fertility was higher, lower were the amounts of organic fertilizer (7/1) necessary to achieve the greatest mycorrhizal effectiveness and functioning.

The results indicated important aspects for the management of nutrient supply in inoculated plants: 1) effective performance requires a suboptimal nutrient supply, 2) higher nutrient supplies, i.e., similar to those applied for optimal growth under non-inoculation conditions inhibit or diminish mycorrhizal performance, and 3) very low supplies also limit the effectiveness of inoculation.

Therefore, inoculation of the "efficient" isolate for a given soil condition may not achieve effective mycorrhizal performance and may not result in positive responses in crop growth and yield, depending on the supply and availability of nutrients (Rivera et al., 2003).

The second group of experiments was conducted in microplots (1 m^2) with different crops in two types of soils and each experiment was repeated three times (Ruiz et al., 2012). Six AMF isolates were evaluated and inoculated by applying between 20-50 g $plant^{-1}$ according to Sieverding (1991). The results were similar since a differentiated and reproducible response to inoculation was found ($p<0.05$), but with the characteristic that in each type of soil the greatest responses in yield (Table 2), colonization and macronutrient concentrations (NPK) in different crops were achieved with the same isolate, in this case *G. intraradices*.

In the experiments carried out in Eutric Nitisols, the efficient inoculated isolate was *G. mosseae*, i.e., when changing soil type, another isolate was applied with which all crops showed the greatest response. Therefore, the efficient isolate to be inoculated will depend on the soil type and under the appropriate soil condition it behaved as the efficient isolate for the group of crops studied.

In these experiments it was also found that certain crops also presented an effective response to inoculation with another isolate. This behaviour was denominated culture-AMF isolate compatibility, for example the cassava crop with *G. clarum* isolate (Table 2); however, this compatibility did not prevent from also establishing an equally effective symbiosis with the inoculation of *G. intraradices,* which was the "efficient" isolate for all crops in this edaphic condition (Rivera et al., 2007).

Likewise, the table shows examples of the diversity of response types of crops, some of which only showed large response values to the application of one isolate, such as taro, and others with a high response to several isolates, such as sweet potato, but in all cases an isolate was found whose inoculation presented the greatest effectiveness in each of the cultures. In the pepper and cucumber crops, the experiments included other isolates in the comparison and although the inoculation of *G. intraradices* maintained its high effectiveness, it was found that another isolate showed similar results to *G. intraradices*, indicating that there may be more than one effective inoculants for the same soil condition.

Table 2. Effectiveness (EI %) of inoculated AMF isolates for different crops in Eutric Cambisols soils

AMF	Potato	Cassava	Sweet potato	Taro	Yam	Banana	Pepper	Cucumber
G. intraradices	43. 9a	48.8a	397.6a	110.0a	47.8a	68.0a	77.7 a	74.2 a
G. fasciculatum	31.2ab	27.4bc	319.5b	6.6bc	39.8b	56.3a	38.2b	44.5b
G. mosseae	24.7bc	1.1d	186.5c	20.0b	29.5c	10.5cd	37.6b	9.4c
G. clarum	18.0bc	38.0a	7.3d	3.3bc	35.4bc	29.2bc	26.1c	18.7c
A. scrobiculata	1.8 d	20.2c	0.0d	10.0c	17.7d	45.2ab	n.e.	n.e.
G. spurcum	n.e.	n.e.	n.e.	n.e.	n.e.	n.e.	73.1a	71.2a
cv %	12.8	7.1	6.9	8.6	3.5	12.6	4.95	11.31

EI (%) = (inoculated treatment yield-control yield) x100/control yield (Janos, 2007). Different letters in each column imply significant differences ($p<0.05$) between the effectiveness of the applied AMF. n.e. not evaluated. The original designation of the AMF species used was maintained. Table elaborated from Ruiz et al., 2012.

In 2003, a new experimental plan was started under field conditions to compare the effectiveness of three "simple" mycorrhizal inoculants in a wide group of crops and 12 different edaphic conditions (Table 1), with the objective of verifying and extending the previous results, but also to establish if any of soil properties explained the change in the inoculant effectiveness. For these 42 experiments inoculants were prepared according to Fernandez et al., (2000) and the isolates used were: *Glomus cubense* [(Rodriguez et al., 2011)/INCAM-4, DAOM241198], *Rhizoglomus irregulare* [(Sieverding, et al., 2014) Syn. *Rhizophagusirregularis*/INCAM-11, DAOM711363] and *Funneliformis mosseae* [(Schüßler and Walker, 2011)/INCAM-2]. Inoculants were identified according to the isolate they contained.Inoculation was performed via seed coating (Rivera and Fernandez,2006; Ruiz et al., 2012), with the exception of coffee plant seedlings, vitroplant adaptation and tobacco seedbed in which the inoculant was applied under the seeds or vitroplants, prior to sowing or transplanting.

In all cases a positive response of yields to inoculation was found, always with significant differences ($P<0.05$) with the non-inoculated control, as well as with significant differences (P<0.05)between the yields or biomass productions caused by the different inoculants. Table 3 shows the effective inoculant for each crop-soil combination. It can be observed that, for the same edaphic condition, the inoculant that produced the greatest response was the same for different crops studied and that each of the inoculants was effective in a range of edaphic conditions that included different types of soils or soil management. Also, the intensity of the response achieved with the inoculants depended on the crops, soil fertility and nutrient supply. This "generalist" behavior seems to be a consequence of the type of isolates used for the formulation of these inoculants and even several authors (Helgason and Fitter, 2009; Van der Heidjen et al., 2015) refer to *R. irregulare* and *F. mosseae* species as generalists with plant species.

Table 4 presents the ranges of different soil properties that were associated with the edaphic conditions in which each inoculant reached its highest effectiveness. With the exception of pH-H_2O, the rest of the properties presented a high overlapping of ranges in which inoculants were effective, and it does not appear that these properties determined the change in effectiveness of inoculants.

Table 3. Effective inoculant found in different experiments carried out in the period of 2003-2018

Soils / Crops	Vertic Phaozem	Eutric Cambisol	Gley Pelic Vertisol	Albic Pisoplintic Plintosol flooded	Lixic Eutric Rhodic Nitisol	Albic Pisoplintic Plintosol + 2 t/ha $CaCO_3$	Rhodic. Alumic Acrisol	Eutric Cambisol and Fluvisol	Dystric Cambisol/	Albic Pisoplintic Plintosol
Cassava		*R. irre.* (2, 3)			*G. cub.*(3)			*G. cub.*(20)		
Sweet potato		*R. irre.*(2,4, 25)								
Taro		*R. irre.*(2)								
Yam		*R. irre.*(2)								
Banana		*R. irre.*(5)								
Corn					*G. cub.*(4)					*F. moss.*(4)
Beans		*R. irre.*(4)			*G. cub.*(4)			*G. cub.* (4)		
Rice				*R. irre.*(11)						
Tomato		*R. irre.*(6)			*G. cub* (12)					
Coffee nursery					*G. cub.* (13)			*G. cub.*(21)		
Advocate					*G. cub.* (14)					
Tobacco							*G. cub.*(18)			
Jack bean	*R. irre.*(1)	*R. irre.*(4)	*R. irre.*(8)		*G. cub.*(15)		*G.cub.*(19)	*G.cub.*(22)	*F. moss.* y *G.cub.* (23).	*F. moss.*(24)
Urochloa species			*R. irre.*(9)		*G. cub.*(9)	*G. cub.*(17)				*F. moss.*(9)
M. maximus			*R. irre* (9).		*G. cub.*(9,16)					*F. moss*(9)
Forage legumes			*R. irre.*(10)		*G. cub.*(10)					*F. moss.*(10)
Caoba nursery		*R. irre.*(7)						*G. cub.* (7)		

R. irre.: INCAM-11 based inoculant; *G. cub.*: INCAM-4 based inoculant; *F. moss.*: INCAM-2 based inoculant. Sources used to prepare the table: (1): Simó etal. (2019); (2): Ruiz et al.(2012); (3):Joao et al.(2016); (4) : Rivera et al. (2017); (5): Simó et al.(2017); (6): Simó et al. (2018); (7):Falcón et al.(2013); (8): Tamayo et al. (2015); (9): González et al. (2015 a); (10): González et al. (2016); (11) Ruiz et al. (2016); (12) data not published; (13):Rivera et al. (2007); (14): Fundora et al. (2011); (15): Martin et al. (2010); (16): Rosales et al. (2017); (17): Ramirez et al. (2016); (18): data not published; (19): García et al. (2017); (20)and (21) data not published; (22): Bustamante et al. (2010); (23) : Ojeda et al. (2019); (24): Martin et al.(2015).

Table 4. Ranges of different soil properties in which each of the inoculants produced the highest crop yield responses

Inoculants	pH-H_2O	MO g kg^{-1}	Ca	Mg	K	P
			cmol$_c$ kg^{-1}			mg kg^{-1}
INCAM-2	4.7-5.7	17.8-24.1	3.54-7.03	1.2-2.17	0.1-0.62	2.4-3.25
INCAM-4	5.6-7.1	13.6-30.2	4.4-29.4	1.1-6.01	0.13-0.57	1.8 – 89.6
INCAM-11	7.0 – 8.0	25.0-46.5	6.0-44.9	1.4-6.86	0.1-0.62	1.7-14.82

The pH-H_2O showed a well differentiated behavior and it was possible to associate specific pH ranges in which only one of the inoculants was most effective. A slight overlap was found in what can be considered a transition zone of the effective inoculant condition, indicating that the changes in effectiveness were directly related to the reaction value of the soil or substrate in which the mycorrhizal roots were located.

Taking into account the diversity of crops, soil conditions and nutrient supply in the different experiments, and in order to compare the results of experiments, the relative yield response (Janos, 2007) obtained by each inoculant in relation to the non-inoculated counterpart (EI) was initially calculated. Subsequently, a Relative Efficiency Index (REI %) was calculated for each inoculant in each soil-crop combination, considering the maximum response obtained as 100% and the other two responses were expressed as percentages of this. In this way, in each soil-crop combination, the inoculant that originated the highest response reached the value of 100%. From the above, Figure 1 was constructed by ordering each of the soil x crop combinations in increasing order of pH-H_2O and it can be observed how the changes in the effective inoculant condition are directly associated with pH.

Also, there were some new cases of "compatibility" in these cases associated with *Pueraria phaseoloides* cv. Kudzú, *Centosema pubescens* cv. and *Urochloa brizantha* with *G. cubense*/INCAM-4 and *R. irregulare*/ INCAM-11 and tomato with *F. mosseae*/INCAM-2, however the REI achieved in these cases were not higher than those obtained by these crops with the "effective" inoculant.

Figure 1 also shows that beyond the pH zone in which each of the inoculants was effective, there was no regularity in the degree of relative response (REI) that they caused, possibly indicating in these cases the influence of crops and context-dependency on responses.

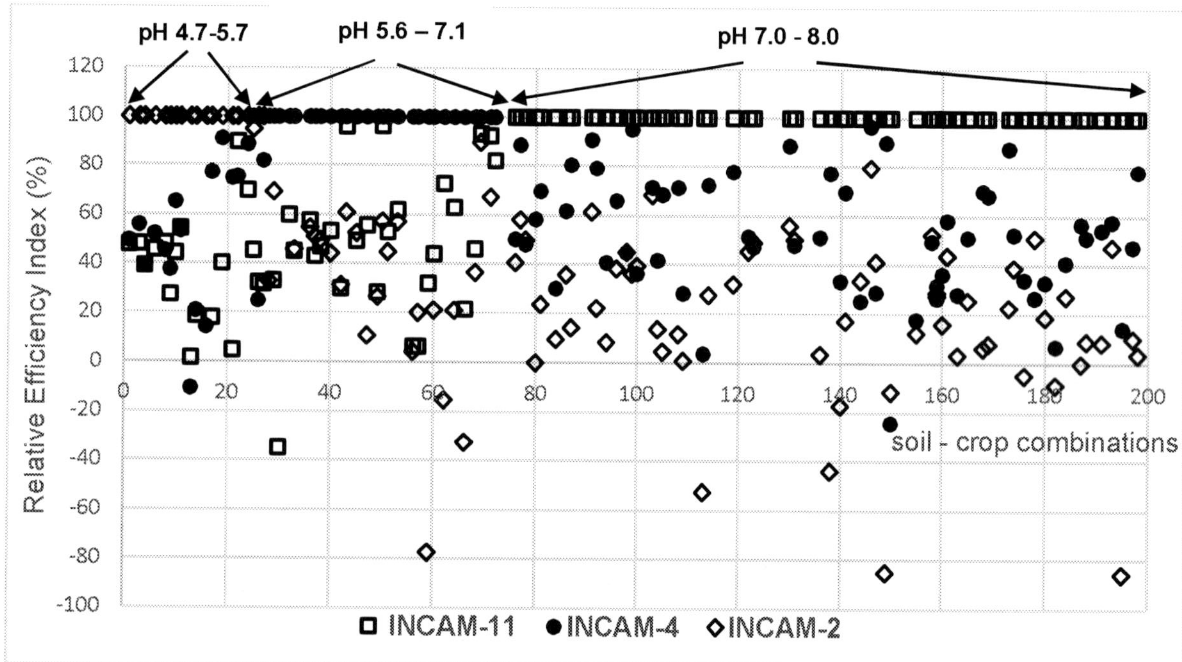

Figure 1. Relative efficiency indices (REI) obtained for three AMF inoculants in each of the different soil-crops combinations. Experiments carried out in the period 2003-2018. In each combination, a positive effect to inoculation was found ($p<0.05$). Combinations are ordered according to increasing pH-H2O values and the pH-H2O ranges associated to the condition of greater effectivity of each inoculant are shown. *F. mosseae* / INCAM-2 pH 4.7 - 5.7; *G. cubense*/INCAM-4 pH 5.6 to 7.1; *R. irregulare*/ INCAM-11 pH 7.0 to 8.0. Sources used to prepare figure the same experimental data used to prepare Table 3. Relative Efficiency Index was calculated for each inoculant in each soil-crop combination, considering the maximum relative response obtained as 100% and the other two responses were expressed as percentages of this.

The information obtained with the change in inoculant effectiveness in the Albic Pisoplintic Plintosol soils when limed (Table 3) in which with an initial pH-H_2O of 4.7 to 4.9 the most effective inoculated isolate for *Uroclhoa* "mulato" was *F. mosseae*/INCAM-2 and that when these soils received doses of 2 to 4 t ha^{-1} of $CaCO_3$ and the pH rose to 6.2 then the highest effectiveness was achieved with the inoculation of *G. cubense*/INCAM-4 and that when limed with 6 t ha^{-1} and the pH rises to 7.2 the most effective isolate resulted to be *R. irregulare*/INCAM-11 (Ramírez et al., 2016), not only make clear the importance of pH-H_2O, but also broaden the scope of inoculant recommendations for amended soils. Similarly, in this same type of soil and growing rice under waterlogged conditions the pH-H_2O is increased to 7.2 and the most effective inoculant turns out to be INCAM-11 (Ruiz et al., 2016).

The colonization percentages always reflected a significant response to inoculation (Figure 2), also showing differences between each of the inoculants, such that in the pH range where the inoculant originated the highest yield responses (Figure 1), it also presented the highest colonization percentages.

In general, the colonization percentages associated with the effective inoculants were found to be in the range of 60 to 70%, although they may depend on several factors such as:

1) The different degree of mycorrhizal dependence and the responsiveness of crops to inoculation, both inter- and intraspecific (Janos, 2007; Chu et al., 2013; Mensah et al., 2015; Maiti and Topo, 2019),
2) The fact that a plant is not colonized with only one AMF isolate (Pellegrino et al., 2012; Van der Heidjen et al., 2015), which originates different colonization percentages based on the abundance of resident mycorrhizal fungi
3) As well as whether nutrient supply was found limiting mycorrhizal functioning (Rivera et al., 2003) among other factors.

However, for each soil- crop combination, the information indicated that the degree of yield response to inoculation of the isolates was directly associated with the colonization percentages obtained (Figure 1; Figure 2) and coinciding with the more general criteria that state that the colonization percentage is an adequate indicator of mycorrhizal performance (Azcón et al., 2003; Lekbert and Koide, 2005; Ortas et al., 2012, Chu et al., 2013; Pellegrino et al., 2015).

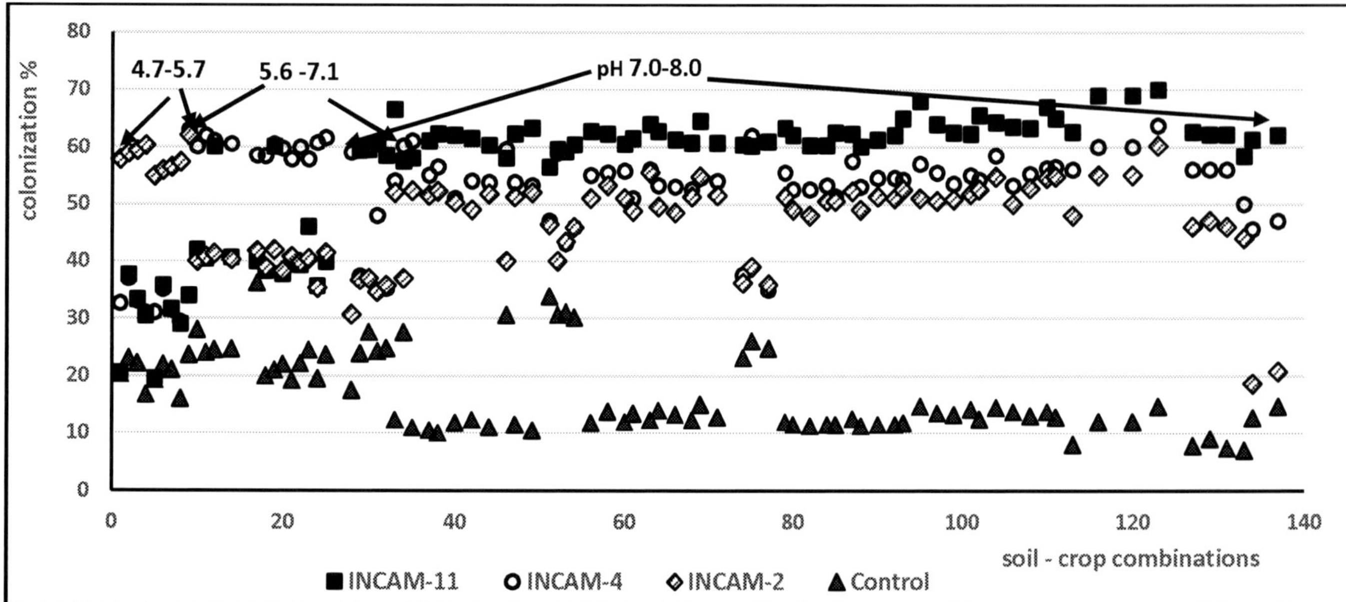

Sources used to prepare figure: González et al. (2015); González et al. (2016); Martin et al. (2010); Martin et al. (2015); Ojeda et al.(2019); Ramírez et al. (2016); Rivera et al. (2017); Rosales et al. (2017); Ruiz et al. (2012);Ruiz et al. (2016); Simó et al.(2017); Simó et al. (2018);Simó et al. (2019) and data not published.

Figure 2. Colonization frequencies associated with the application of three AMF inoculants and the control (non-inoculated counterpart) in each soil-crop combination. Combinations are ordered according to increasing pH-H2O values and the pH-H2O ranges in which each inoculant presented the highest colonization percentages are shown. *F. mosseae*/INCAM-2 from 4.5 to 5.7; *G. cubense*/INCAM-4 from 5.6 to 7.1: *R. irregulare*/ INCAM-11 from 7.0 to 8.0. The confidence intervals at P 95 % in the ANOVA executed in each combination do not exceed 1.5%, therefore differences ≥ 3.0 between the colonization percentages in each soil-crop combination were significant.

The results are also in correspondence with some publications that establish the importance of pH in the effectiveness of inoculated isolates (Mosse, 1973; Hayman and Tavares, 1985; Ruiz et al., 1999; Ouzounidou et al., 2015), which is not surprising, not only because they are edaphic microorganisms, but also because an important part of the mycorrhizal structures are found in the soil. The fact that we have been able to evaluate the effectiveness of the same inoculants in a wide group of crops and soil types, it allowed us to find regularities in the changes in inoculant effectiveness and mycorrhizal performance that they cause when soil conditions and crops varied.

We have found a working model for these inoculants, formulated with generalist isolates with plant species, in which their effectiveness will be directly related to the reaction value (pH-H_2O) of the soil or substrate in which mycorrhization develops. In the appropriate edaphic environment, the inoculant originates an effective symbiosis with any crop dependent on mycorrhization, although the magnitude of the response will depend on the crop (Janos, 2007), on the availability of nutrients in the soil (Rivera et al., 2003) and on whether the community of resident mycorrhizal fungi allows the establishment of an effective inoculation (Dodd and Thompson, 1994), among other factors.

There are also compatibilities between inoculants and crops, but these do not prevent the establishment of an equally effective symbiosis when inoculating the efficient AMF isolate for that edaphic condition; therefore, the criteria for recommending inoculants based on soil pH-H_2O do not vary in order to obtain the benefits of mycorrhization in crops.

Although we cannot ensure that the reproducible effects found are due to effective mycorrhization with the applied isolate or to interaction with resident arbuscular mycorrhizal fungi (Rodriguez and Sanders, 2015), the fact that these reproducible effects have been obtained by integrating the results of experiments performed in 12 different edaphic conditions and with about 20 crops and where the only common thing have been the three inoculants suggests the validity of the direct effect, which is also in correspondence with different reports that under conditions of response to inoculation in non-sterile conditions and using molecular techniques the applied isolates colonizing the crops have been identified (Pellegrino et al., 2012;Rodriguez et al., 2021). In any way, they allow us to use these results as a basis for recommend, which isolate to inoculate.

The scope of the results includes most of the agricultural soils of the country (pH 4.7 to 8), but also incorporates the more acidic soils once they are amended. This has a high practical value as it allows the use of the same inoculant (isolate) for different mycorrhizal-dependent crops in the appropriate pH-H_2O range.

It is important to point out that the criteria obtained for the recommendation of inoculants should not be valid for all AMF species, since according to Opik and Mora (2012), the type of association that each one establishes places them between high and low specificity with crops and also being able to be specific or not with the soil condition. This does not detract from the value of the result obtained, but rather establishes its scope.

4. Integrated Strategy Efficient AMF Inoculants and Fertilization

From the beginning of the work programme, the benefits of mycorrhizal inoculation in reducing the quantities of fertilizers needed to achieve high yields and satisfactory nutritional status of crops were sought, with the aim of designing more economical technologies with less use of external inputs. The first works were related to organic fertilizer applications for the production of coffee plant seedlings and were discussed in the previous topic.

Subsequently, “efficient” isolates per soil type were inoculated with different doses of NPK mineral fertilizers in several crops and in two types of soils. Table 5 summarizes the results of different field experiments all conducted with a similar experimental scheme and evaluating five fertilizer doses (percentages of the recommended NPK dose for each crop: 0, 25, 50, 50, 75 and 100%) with and without inoculation of the “efficient” isolate, in Eutric Cambisols soils (Ruiz et al., 2012). In this case inoculation was realized applying between 20 and 50 g plant-^{1} according to Sieverding (1991).

In all crops, an increasing response to fertilization was found, as well as a positive response to inoculation and in all cases the inoculated crops required lower amounts of fertilizer to obtain high yields similar to those obtained when applying the 100% NPK dose in non-inoculated crops. Likewise, a high reproducibility of the results was found in the two replications carried out.

The decreases in fertilizer dose for each crop were found to be between 25 and 75%, the most common being a 50% decrease in fertilizer amounts, although the magnitude of the effect will depend on crops through their degree of response to inoculation (Janos, 2007), the requirements and dynamics of nutrient absorption by each crop and the availability of the different macronutrients in the soil (Siqueira et al. 1993; Rivera et al., 2003).

Crop colonization percentages increased significantly with inoculation and values between 60 and 70% were reached in the inoculated treatments that received the fertilizer doses with which the highest response to inoculation was achieved, indicating that the higher mycorrhizal functioning achieved explains the decrease in the amounts of fertilizer needed to obtain high yields. Higher fertilizer doses decrease colonization percentages. Consistent with decreases in fertilizer requirements, inoculation always increased fertilizer use efficiency (EFU) when compared to the treatment that received 100% mineral fertilization.

However, in inoculated treatments, the application of fertilizer dose lower or higher than those that caused the greatest response, decreased the colonization percentages and the response to inoculation, so that depending on the availability or supply of nutrients, the inoculation of the "efficient" isolate may not be effective (Rivera and Fernandez, 2007).

Improvement in plant nutrition by increasing nutrient uptake capacity is one of the most reported benefits of mycorrhization and while it has been associated primarily with P and N uptake (Smith and Smith, 2011; Bagyaraj et al, 2015; Hodge and Storer, 2015), mycorrhizal involvement in the uptake of virtually all essential nutrients has been demonstrated (Marschner and Dell, 1994; George, 2000; Liu et al., 2002; Pellegrino et al., 2015; Bhandari and Garg, 2017).

Increased nutrient uptake capacity leads not only to more efficient uptake of fertilizers, but also of nutrients in the soil which is expressed by decreasing critical levels of elements in the soil (Siqueira and Franco 1988; Gonzalez et al., 2018), as well as reducing nutrient leaching losses (Ashgari and Cavagnaro, 2014; Cavagnaro et al., 2015); however although positive results associated with inoculation are increasing (Pellegrino et al., 2015; Zhang et al., 2018; Schutz et al., 2018), results integrating inoculant management with fertilization schemes and crop technologies are still scarce and will be necessary to demonstrate the benefits and feasibility of mycorrhizal inoculants in agricultural production (Hamel and Strullu, 2006; Ryan et al., 2019).

Similar results with this same experimental scheme in Eutric Nitisol soils but inoculating *G. cubense*/INCAM-4, an "efficient" AMF isolate for this soil type (pH-H2O: 6-7), were found in cassava, maize and bean crops (Joao et al., 2017; Morejon et al., 2017; Rivera et al., 2017).

Table 5. NPK fertilizer dosage, associated colonization percentage and efficiency in fertilizer utilization (EFU) for different crops inoculated with *R. irregulare*/INCAM-11 in Eutric Cambisols soils. Results of field experiments

Crop	Treatment	First-year			Second-year		
		t ha^{-1}	Col.%	EFU kg kg^{-1}	t ha^{-1}	Col.%	EFUkg kg^{-1}
Cassava	Control	35,5 b	18 b		37,1 b	20 c	
	AMF+ 25% NPK	48,8 a	60 a	129,8	52, 5 a	70 a	145,4
	100% NPK	50,3 a	23 b	36,1	52,0 a	28 b	36,4
	$SE_{\bar{x}}$	1,26**	1,1**		1,05**	0,9**	
Sweet potato	Control	18,8 b	7 c		27, 2 b	12 c	
	AMF + 50%NPK	36,1 a	70 a	129,1	40,1 a	80 a	96,3
	100% NPK	38,3 a	35 b	72,8	41,4 a	26 b	53,0
	$SE_{\bar{x}}$	0,91**	1,15**		1,32*	0,9**	
Taro Colocasia	Control	25,5 c	7 c		29,0	16 c	
	AMF + 50% NPK	36,7 a	63 a	36,7	40,9 a	76 a	39,0
	100% NPK	38,3 a	21 b	21,0	41,4 a	16 b	20,3
	$SE_{\bar{x}}$	1,00**	0,85**		1,08**	0,85**	
Taro Xanthosoma	Control	17,9 c	7 c		23,2 d	14 b	
	AMF + 50% NPK	40,1 a	61 a	144,2	42,7 a	75 a	126,6
	100% NPK	40,3 a	22 b	72,7	43,2 a	17 b	64,9
	$SE_{\bar{x}}$	0,71**	0,9**		1,19**	1,2*	
Yam	Control	17,8 c	12 c		25,1 d	20 c	
	AMF + 50% NPK	26,9 a	68 a	50,6	33,0 a	79 a	43,9
	100% NPK	27,9 a	35 b	28,1	32,8 a	26 b	21,4
	$SE_{\bar{x}}$	0,83**	1,2**		1,04	1,5**	

Table elaborated from Rivera and Fernandez (2003) and Ruiz et al. (2012).

Crop	Treatment	First-year			Second-year		
		t ha $^{-1}$	Col. %	EFU kg kg^{-1}	t ha^{-1}	Col. %	EFU kg kg^{-1}
Cucumber	Control	5,9 c	9,0 c		7,0 c	12,0 c	
	AMF + 50% NPK	13,1 a	68,0 a	61,5	15,4 a	72,0 a	72,1
	100% NPK	13,35 a	33,0 b	31,8	15,6 a	35,0 b	36,6
	$SE_{\bar{x}}$	1,0**	0,31**		0,61**	0,46**	
Garlic	Control	7,89 c	5,0 c		8,35 c	7,0 c	
	AMF + 50% NPK	14,87 a	73,0 a	50,76	17,01 a	78,0 a	62,98
	100% NPK	15,22 a	25,0 b	26,65	17,50 a	28,0 b	33,27
	$SE_{\bar{x}}$	0,87 **	0,80 **		0,79 **	1,80 **	
Potato	Control	18,0 d	10 c		20,5 d	20 b	
	AMF +75% NPK	29,2 a	65 a	31,8	36,8 a	78 a	46,2
	100% NPK	30,1 a	28 b	25,7	37,2 a	20 b	35,5
	$SE_{\bar{x}}$	0,97**	0,95**		1,46**	1,05**	
Tomato	Control	18,6 e	12,0 c		20,1 e	14,0 c	
	AMF + 75%NPK	40,9 a	72,0 a	116,6	41,9 a	73,0 a	114,0
	100% NPK	41,1 a	36,0 b	88,2	42,0 a	33,0 b	85,9
	$SE_{\bar{x}}$	0,8**	2,13 **		0,76**	0,76**	
Banana	Control	15,6 d	16,5 b		12,3 d	11,0 b	
	AMF +75% NPK	29,6 a	76,0 a	28,2	25,6 a	73,0 a	26,8
	100% NPK	29,7 a	15,0 b	21,3	25,0 a	10,0 b	19,2
	$SE_{\bar{x}}$	0,72**	0,68**		0,54**	0,70**	0,72

Col.: Colonization. Different letters in each column for the same crop indicate significant differences ($p<0.05$) according to Duncan's test. In each crop, the AMF+ fertilizer treatment was the one that presented a similar yield at 100% NPK and with the lowest dose of fertilizer. The control treatment did not receive any fertilizer. pH-H_2O: 7.3 - 7.6.

$$\textbf{EFU}\ (\text{kg kg}^{-1}) = \frac{[\text{Yield selected Treatment (t ha}^{-1}) - \text{Yield without fertilizers (t ha}^{-1})] \times 1000}{\Sigma \text{kg ha}^{-1} \text{ applied of N+P+K}}$$

Since 2003, in order to establish the possible differentiated effects of inoculants on the amounts of each macronutrient to be supplied and to better guide the fertilization of crops, about 25 experiments were carried out in which 4 to 5 levels of each macronutrient (between 0 and 100% or 125% of the recommended dose of the nutrient for high yield) were evaluated in the presence or not of the effective inoculant for each soil condition and with a fixed background of the other two macronutrients, which in most cases coincided with 100% of the recommended dose for each one. Inoculation was performed via seed coating. The crops evaluated were cassava, sweet potato, banana, corn and forage grass species, and the experiments were carried out in Eutric Cambisols, Eutric Nitisols, Gley Pelic Vertisol and Albic Pisoplintic Plintosol soils (González, 2014; Ruiz et al., 2016 a, 2016 b; Espinosa et al., 2015; 2016, 2017; Rivera et al., 2017).

In the inoculation absence, all crops showed a positive response to the application of each macronutrient, with the highest yields associated with high doses of fertilizers, indicating that the experiments were conducted under conditions of response to fertilization and therefore that the soils themselves do not guarantee the necessary nutrients for the crops. The exception to this behavior were the experiments on Gley Pelic Vertisol soils with different *Urochloa* species in which no response to P doses was found, although a positive response was found in these soils to the application of the other two macronutrients.

In all crops and soil types a positive response to inoculation was found, expressed in that high yields similar or higher than those obtained when the crops received 100% of the fertilization, but with lower doses of fertilizers. Even in the Gley Pelic Vertisol soil in which although no response to phosphorus fertilization was found, however there was a response to the inoculation expressed by higher yields, colonization percentages and higher concentrations of N and K in the biomass of the grass species (González, 2014), indicating the benefits in the nitrogen and potassium nutrition of the grasses.

As examples, the results obtained on cassava and brachiaria grasses (*U. decumbens*) both growth on Eutric Nitisol soils are presented. In cassava crop (Figure 3) a significant and increasing response to the applications of the three macronutrients was found and in the absence of the inoculation the highest yields were associated with the highest doses studied, which coincide with the NPK fertilization recommendations for this crop (MINAG, 2012).

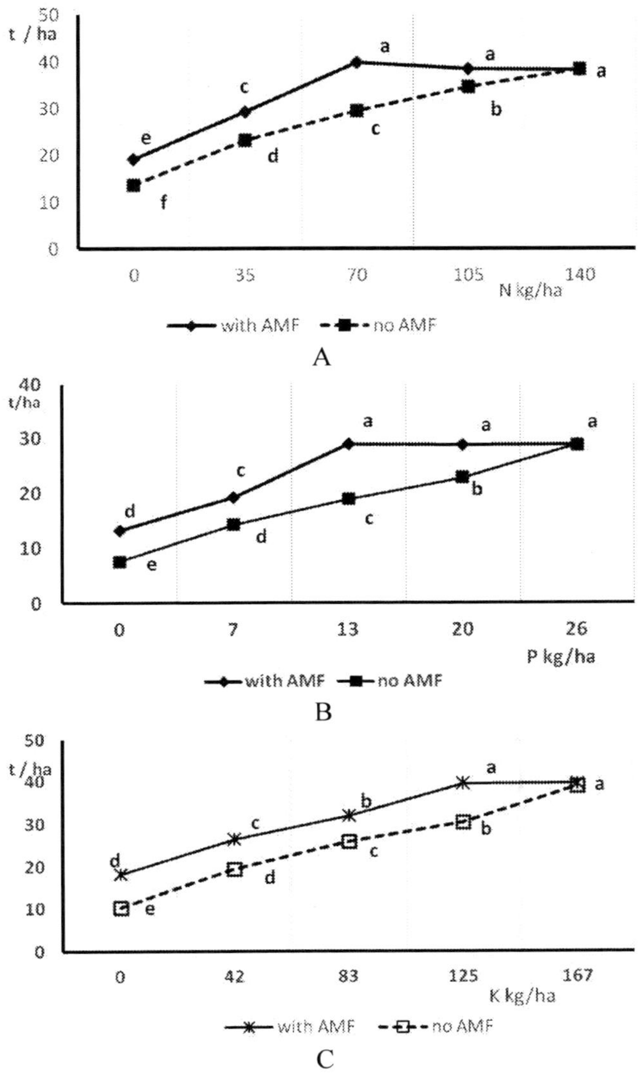

Figure 3. Response to *R. irregulare*/INCAM-11 inoculation and doses of three macronutrients in *Manihot esculenta*`Señorita´on Eutric Cambisol soils (Ruiz *et al.*, 2016 a). A) Nitrogen dose-response curve; B) P dose-response curve and C) K dose-response curves. Fixed backgrounds: 140, 26 and 167 kg ha^{-1} of N, P and K. Different letters imply significant differences at P<0.05 by Tukey's test. A) SE $_{\bar{x}}$=1.05; B) SE$_{\bar{x}}$= 1.02; C) SE$_{\bar{x}}$=1.1. Inoculation via stake tip coating.

In the presence of mycorrhizal inoculation the fertilizer requirements decreased to 50% of N and P and 75% of K. Corresponding to these doses, the highest values of colonization percentages were reached, which ranged between 70 and 75%, of mycorrhizal spores which were found between 460 and 620 spores in 50 g (Figure 4) as well as the highest coefficients of fertilizer utilization (Table 6).

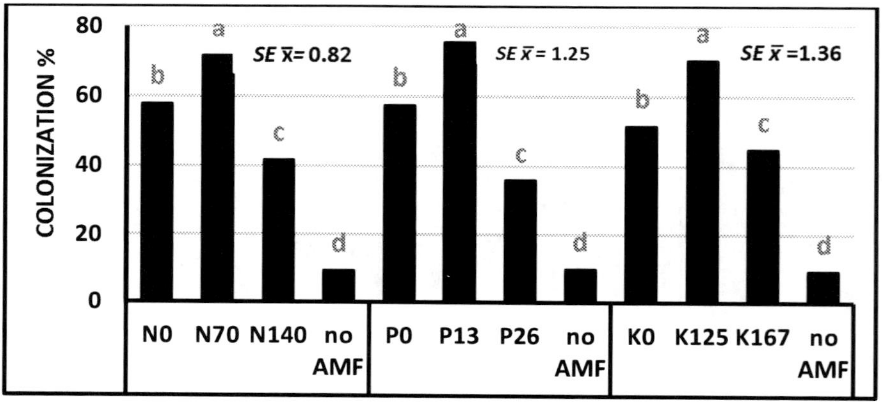

A

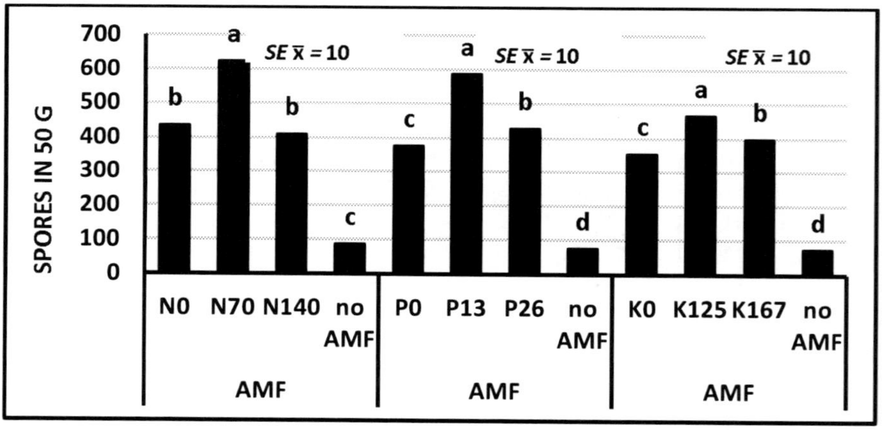

B

Figure 4. Percentages of colonization (A) and spores (B) in selected treatments of the N, P and K experiments on cassava cultivar "Señorita," in Eutric Cambisol soil (Ruiz et al., 2016 a). N-70+AMF; P-13+AMF and K-125+AMF are the fertilized treatments with the highest yield response to inoculation in each experiment. The column no AMF is the average of all non-inoculated treatments in each experiment. All treatments received a fixed background of the other two macronutrients of 140, 26 and 167 kg ha-[1] of N, P and K respectively. SE $_{\bar{x}}$: Standard error.

The information obtained links the effects obtained on yield and fertilizer dose reduction with a higher mycorrhizal performance, given not only by the higher colonization percentages obtained, but also by the greater quantities of spores reproduced by the inoculated crop.

Table 6. Effect of inoculation with *R. irregulare*/INCAM-11 on the "Apparent recovery efficiency" (ARE) by the cassava cultivar, "Señorita," in Eutric Cambisol soil

Nitrogen Experiment		Phosphorus Experiment		Potassium Experiment	
Treatment kg ha^{-1}	ARE-N %	Treatment kg ha^{-1}	ARE-P %	Treatment (kg ha^{-1})	ARE-K (%)
N0-PK		P0-NK		K0.NP	
N70-PK +AMF	54	P13-NK+AMF	43	K125-P+AMF	47
N140-PK	28	P26-NK	25	K167-NP	35

(Table elaborated from Ruiz et al., 2016 a).

N70-PK+AMF; P13-NK+AMF and K125-NP +AMF are the fertilized treatments with the highest response to inoculation in each experiment. N-140; P-26 and K-167 are optimal supply of each macronutrient (kg ha^{-1}) for non-inoculated crops. ARE= (increase in extraction of a nutrient in relation to the control) x 100/dose of the nutrient applied (Stewart, 2007). Fixed backgrounds: using 140, 26 and 167 kg ha-^{1} of N, P and K respectively.

Spore reproduction has not always been associated with effective mycorrhizal functioning and is highly dependent on the context in which mycorrhization takes place, as well as on crops (Bever et al., 1994); however Willis et al. (2013) propose that the C gains associated with effective mycorrhizal functioning can be used in producing spores and on the other hand Helgason et al. (2007) and Van der Heidjen et al. (2015) consider that generalist species are characterized by high spore production, therefore it is not surprising that treatments with higher functioning and with higher response to inoculation present higher amounts of spores and vice versa.

Experiments with *Urochloa* species for forage with six annual cuttings and grown on Eutric Nitisols soils (Table 7) presented positive responses to applications of the three macronutrients and the highest biomass yields were found with annual doses of 300, 13 and 83 kg ha^{-1} of N, P and K respectively in the first four years, increasing the annual potassium requirement to 125 kg ha^{-1} for the last two years.

In the presence of inoculation with *G. cubense*/INCAM-4, similar high yields were obtained, but with lower annual doses of 200 and 42 kg ha^{-1} of N and K respectively, and for the last two years the dose of potassium fertilizer was increased to 83 kg ha^{-1} year^{-1}; it was not necessary to apply phosphoric fertilizer in any of the years of the experimental period.

Table 7. Effect of inoculation with *G. cubense*/INCAM-4 and mineral fertilization on dry biomass production (t ha^{-1}) of *U. decumbens* grown in Eutric Nitisol soil. Selected treatments of the three experiments

Experiments/Treatments	Year 1	Year 2	Year 3	Year 4	Year 5	Year 6
N and AMF Experiment						
N0-PK	10,81 c	8,69 d	7,90 c	6,89 c	6,56 c	5,34 c
N200-PK	18,13 b	15,56 c	7,90 c	6,89 c	6,56 c	5,34 c
N200-PK + AMF	21,39 a	17,01 b	18,33 a	17,39 a	17,54 a	16,52 a
N300-PK	21,42 a	19,61 a	18,36 a	17,73 a	17,14 a	16,77 a
SE A x B	0,48**	0,47**	0,43**	0,52**	0,51**	0,44**
P and AMF Experiment						
P0-NK	17,95 b	13,25b	15,03 b	14,06 b	13,23 b	11,98 b
P0-NK + AMF	20,96 a	14,71 b	18,89 a	18,05 a	15,98 a	15,12 a
P13.1-NK	21,34 a	17,22 a	19,40 a	18,15 a	16,05 a	15,14 a
SE A x B	0,44**	0,41**	0,43**	0,41**	0,34**	0,36**
K and AMF Experiment						
K0-NP	15,88 c	10,89 d	12,21 c	11,63 c	9,88d	8,39 d
K41.6-NP	18,02 b	13,93 c	14,9 b	13,9 b	12,7 c	12,1 c
K41.6-NP+AMF	20,81 a	15,33 b	18,01 a	17,73 a	15,09 b	14,02 b
K83.3-NP	20,33 a	18,14 a	18,73 a	17,47 a	14,79 b	13,55 b
K83.3-NP + AMF	21,01 a	18,09 a	17,81 a	17,69 a	16,92 a	15,91 a
K125-NP	20,97 a	17,77 a	18,97 a	17,73 a	17,15 a	16,07 a
SE A x B	0,49**	0,42**	0,47**	0,43**	0,45**	0,45**

(Table elaborated from González, 2014).

Different letters in each column for each experiment correspond to significant differences $P<0.05$ according to Duncan's test. In the experiments the interaction A x B was significant. Fixed backgrounds: 200, 22 and 83 kg ha^{-1} per year of N, P and K respectively, with the exception of the last two years when the K background amounted to 125 kg ha^{-1}. Annual re-inoculation started in third year after the last cutting of the low rainfall period with doses equivalent to 15 to 20 kg of inoculant in 200 L of water applied on freshly cut forage and with the first rains.

Inoculation also led to significant increases in the percentages of colonization and spores, with the highest values associated with applications of fertilizers that originated the highest biomass response in each experiment (Gonzalez, 2014), these values were found for *U. decumbens* between 60 and 65% colonization and between 600 to 650 spores in 50 g of soil. Macronutrient concentrations in aboveground biomass were also increased (Table 8) indicating that when inoculation is effective, it causes a greater mycorrhizal performance, which is expressed in a greater absorption of macronutrients from both soil and fertilizers and a satisfactory nutritional status, all of which has repercussions in reaching high yields with lower fertilizer supplements.

Table 8. Effect of inoculation with *G. cubense*/INCAM-4 and mineral fertilization on N, P and K concentrations (g kg^{-1}) in dry biomass of *U. decumbens* grown in Eutric Nitisol soil. Selected treatments of the three experiments

Experiments/Treatments	Year 1	Year 2	Year 3	Year 4	Year 5	Year 6
N and AMF Experiment	N g kg^{-1}					
N0-PK	11,5 c	10,69 d	12,0 c	11,7 c	11,9 c	12,3 c
N200-PK	16,13 b	15,56 c	15.79 b	15,68 b	15,94 b	14,95 a
N200-PK + AMF	18,3 a	17,01 b	17,5 a	18,5 a	18,0 a	17,7 a
N300-PK	17,3 a	19,61 a	18,1 a	18,5 a	17,9 a	18,3 a
SE A x B	0,5**	0,4**	0,4**	0,4**	0,5**	0,3**
P and AMF Experiment	P g kg^{-1}					
P0-NK	1,8 b	1,7 b	1,7 b	1,8 b	1,9 b	1,7 b
P0-NK + AMF	2,2 a	2,1 b	2,1 a	2,3 a	2,3 a	2,0 a
P13.1-NK	2,1 a	2,2 a	2,0 a	2,2 a	2,2 a	2,1 a
SE A x B	0,1**	0,1**	0,1**	0,1**	0,1**	0,1**
K and AMF Experiment	K g kg^{-1}					
K0-NP	12,9 c	12,9 c	13,2 c	1.25 c	13,0 c	1,27 c
K41.6-NP	14,1 b	14,7 b	14,9 b	13,9 b	13,7 c	13,1 c
K41.6-NP+AMF	15,9 a	16,2 a	16,3 a	15,5 a	14,3 b	1,43 b
K83.3-NP	15,7 a	16,1 a	16,5 a	15,6 a	14,7 b	1,45 b
K83.3-NP + AMF	15,8 a	16,5 a	16,4 a	15,7 a	16,3 a	1,66 a
K125-NP	15,9 a	15,8 ab	16,7 a	15,7 a	16,3 a	1,67 a
SE A x B	0,3**	0,3**	0,4**	0,3**	0,5**	0,5**

(Table elaborated from González, 2014).

Different letters in each column for each experiment correspond to significant differences $P<0.05$ according to Duncan's test. In the three experiments the interaction A x B was significant. Fixed backgrounds: 200, 21.8 and 83.3 kg ha^{-1} per year of N, P and K respectively, with the exception of the last two years when the K background amounted to 125 kg ha^{-1}. Annual re-inoculation started from the third year.

An important aspect in this type of crop is to establish the timing, route and dose of re-inoculation. Even though this is a perennial crop, an effective mycorrhizal performance is not maintained throughout the life of the plantation. This can be seen in the disappearance of the inoculation effect in the second year and achieved again from re-inoculation (tables 7 and 8). The re-inoculation frequency seems to depend on the intensity of cuttings, grass species and soil type (González, 2014; Rivera et al., 2020). When *U. decumbens* is grown on Eutric Nitisol soils, re-inoculation is done annually and rises to every two years when *U.* "hybrid" mulato, with a larger root system, is grown. When the annual cuttings go from 6 to 4 or 5, the need to re-inoculate is generally extended by one year. The re-inoculation time coincides with the last cutting of the low rainfall period and when the rains begin.

The high biomass yields of *U. decumbens* (Table 7) and its potassium contents (Table 8) lead to important exports of this element from the system, in the order of 250 to 275 kg ha^{-1} yr^{-1} in both non-inoculated and inoculated treatments; however, the exports were much higher than the fertilizers received and nevertheless the K contents in the biomass of the best treatments have been maintained with little variation between years (Table 8) and explainable by the high participation of soil K in the crop nutrition.

Urochloa species are characterised by their ability to take up significant amounts of potassium from forms hardly available in the soil (Rosolem et al., 2012; Volf et al., 2018). Figure 5 shows the significant decreases over the years in exchangeable potassium, but especially in non-exchangeable potassium (extraction with boiling HNO_3 1M), which were higher in the inoculated treatments. It was estimated that 50-60% of the potassium extracted by the inoculated plants corresponded to hardly available forms. In addition to the importance of taking into account these K forms for the definition of fertilization systems for this crop (Rosolem et al., 2012), it is necessary to know them to ensure the sustainability of biofertilized systems, as well as to incorporate input and export balances and soil and foliar analysis to the fertility monitoring of these systems. This situation, which is more easily detected in perennial crops, has led us to start new experiments on fertility management and sustainable nutrient management in inoculated crop sequences.

A summary of the effects found in different experiments when inoculating the efficient AMF isolate on the requirements of each macronutrient (NPK) is presented in Table 9. Similar to results discussed above, a positive response to inoculation was found in all crops and soil types, such that the amounts of nutrients required to obtain high yields were lower than those required in the absence of inoculation to obtain similar yields, and this effect was found for any of macronutrients, coinciding with the most current criteria that relate the effect of arbuscular mycorrhizae on the absorption of the different essential elements (Marschnery Dell, 1994; Liu et al., 2002; Bhandari and Garg, 2017).

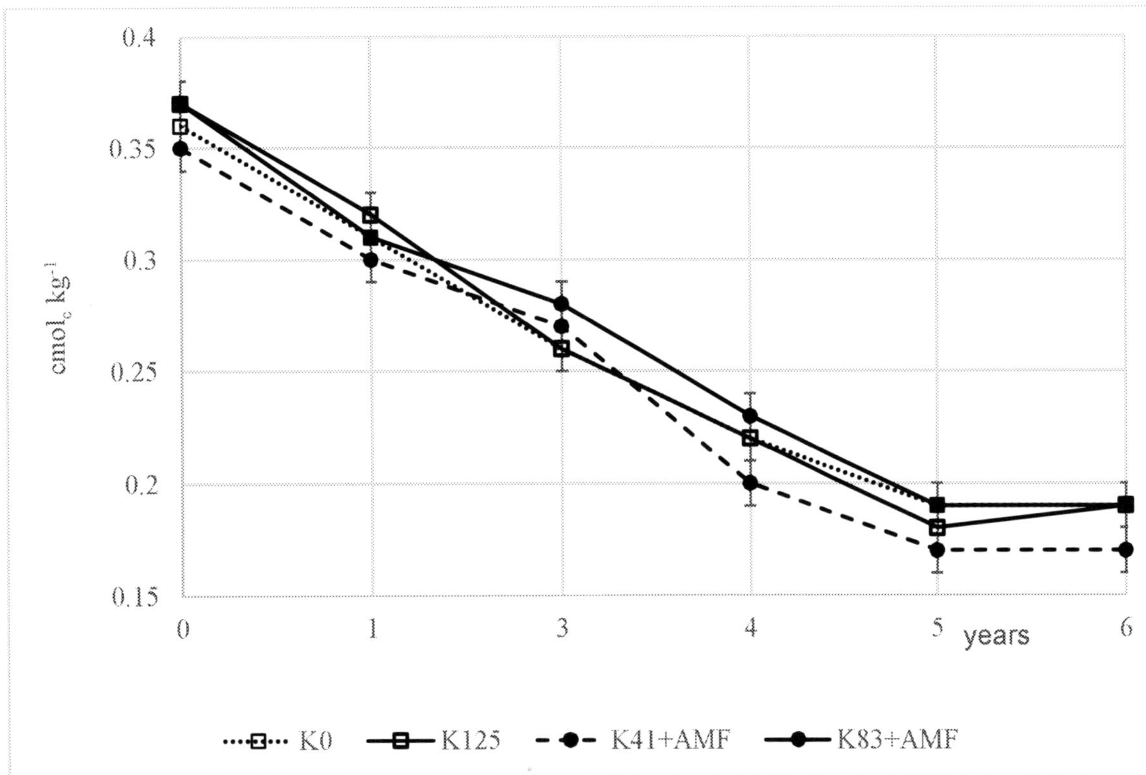

A

Figure 5. (Continued).

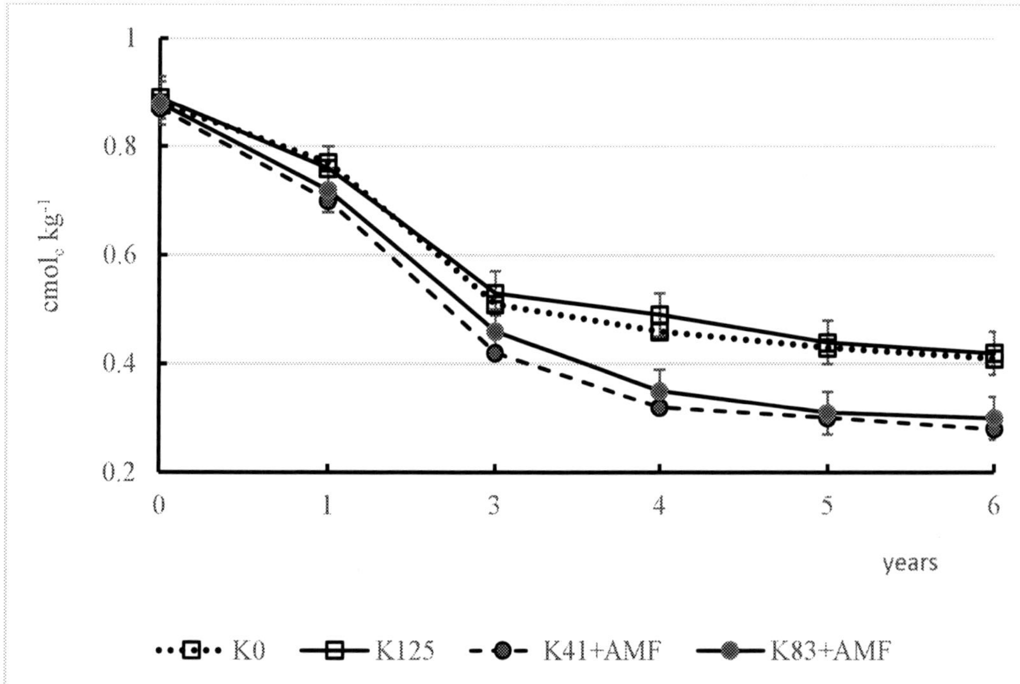

B

Figure 5. Dynamics of exchangeable (A) and non-exchangeable K (B) in the soil over six years. K0 and K125: treatments non inoculated with dose of 0 and 125 kg ha^{-1} yr^{-1} of K respectively; K41-AMF and K83-AMF: treatments inoculated with *G. cubense*/INCAM-4 with doses of 41 and 83 kg ha^{-1} yr^{-1} of K respectively. All treatments received the same fixed funds NP (22 and 83 kg ha^{-1} yr^{-1} respectively). Vertical bars show the confidence interval at P 95 %. (Elaborated from González, 2014).

Table 9. Effect of inoculation with efficient AMF isolates on fertilizer requirements (N, P, K) for high yields in different crops and soil types. Response curves to each macronutrient with and without AMF

Soils/AMF efficient isolates	Crops	Optimum fertilizer doses in each experiment (without AMF inoculation) kg ha^{-1} $year^{-1}$			Fixed funds kg ha^{-1} $year^{-1}$	Suitable fertilizer rates for inoculated crops kg ha^{-1} $year^{-1}$		
		N	P	K	N-P- K	N	P	K
Eutric Cambisol/INCAM-11	Cassava, clon Señorita, two years (1)	140	26	167	140-26-167	70	13	125
	Sweet potato 2 clons, two seasons and two years (2,3,4)	90	33	125	90-33-125	60	22	63
	Banana FHIA18, three cycles (5)	300	33	600	300-33-600	200	11	400
	Banana FHIA21, two cycles (6)	300	50	720	300-50-720	200	25	400
Gley Pelic Vertisol/INCAM-11	Brachiaria two species and two years (7)	300	0	83	200-50-100	200	0	42
Eutric Nitisol/INCAM-4	Brachiaria four species and four years (7,8,9)	300	13	83	200-22-83	200	0	42
	Corn (N experiment) (6)	120			0-30-50	80		
	x. Triticosecale Wittmack (N experiment) (10)	150			0-50-74	100		
Albic Pisoplintic Plintosol/INCAM-2	Corn (N experiment) (6)	150			0-50 - 120	100		
	B. decumbens four years (not published)	200	26	208	150-13-167	150	13	125
	B. "hybrid mulato" (N and cattle manure) (9)	150 kg ha^{-1} of N and 30 t ha^{-1} de manure				105 kg ha^{-1} of N and 15 t ha^{-1} of manure		
	M. maximus cv. Likoni (only except for N - cattle manure (11)	150 kg ha^{-1} of N and 25 t ha^{-1} of manure				105 kg ha^{-1} of N and 12,5 t ha^{-1} of manure		

AMF: The "efficient" isolate was inoculated based on the results of Table 3. Inoculation via seed coating in grains and forage crops (Rivera and Fernandez, 2006) and in tropical roots (Ruiz et al., 2012); in banana, inoculation in the transplant hole (Simó et al., 2015). Re-inoculation in forage crops according to González (2014). In the case of banana, the application of phosphoric fertilizer was carried out only at the beginning. Table elaborated from data published in (1) Ruiz etal. (2016 a); (2) Espinosa et al. (2015); (3) Espinosa et al. (2016); (4) Espinosa et al. (2017); (5) Ruiz et al. (2016 b); (6) Rivera et al. (2017); (7) González (2014); (8) González et al. (2015 a); (9) González et al. (2015 b); (10) Plana et al. (2016); (11) Ramirez et al. (2017).

An interesting aspect derived from these experiments was that in most of the experiments the fixed funds applied coincided with the optimum fertilizer amounts for the crops in the absence of inoculation. That is, the positive response to inoculation was manifested in the presence of a single macronutrient in limiting condition. While it is recognized that plants need to be under suboptimal nutrient supply conditions for mycorrhization to be effective (Johnson et al., 1997; Azcón et al., 2003 Habte, 2006) the information obtained indicates that, at least under these conditions, this suboptimal supply does not have to be manifested with all macronutrients in unison. This behaviour allows a better integration of mycorrhizal inoculants with nutrient supply schemes, especially with those nutrients that crops export in greater quantities, such as potassium, and which may require high doses in inoculated systems to avoid negative imbalances in the soil.

The large number of experimental results with different crops and soil conditions, allows us to suggest that it is feasible to integrate appropriate mycorrhizal inoculants in fertilization and manuring schemes and as part of crop technologies to obtain high yields and an adequate nutritional status, increasing the efficiency of use of mineral or organic fertilizers. The development of these fertilization systems for inoculated crops requires monitoring of soil fertility including the balance of inputs and exports and the nutritional status of the plants in order to direct the supply of nutrients to the mycorrhizal crops efficiently and to achieve in a sustainable way the benefits of a satisfactory mycorrhizal performance over successive crops; benefits that are not only expressed in crop nutrition (Rillig et al., 2019).

5. Integrated Strategy: Efficient AMF Inoculants, Fertilization and Green Manure

Although the previous topics have presented results that support inoculant application based on soil pH and take advantage of their benefits in reducing the amount of fertilizers, the use of mycorrhizal inoculants in agricultural production requires their integration with different cultural practices. Legume-based green manures have a number of benefits for agriculture associated not only with nitrogen inputs and nutrient recycling, but also as soil cover and erosion reduction (Bunch, 2006; Cherr et al., 2006; Vargas et al., 2017) and their use should be more widespread. García et al. (2000, 2002) recommended green manure species for Cuban agriculture and evaluated the decreases in nitrogen fertilizer when using them in different crops. These species are mycotrophic and from the use of inoculants based on generalist isolates, from the positive results found with the effect of the permanence of the inoculant applied to a crop on the mycorrhization of the crop in succession (Rivera and Fernandez, 2003) and the fact that the hyphae of the same AMF species connect the different crops present in the agroecosystem (van der Heijen et al., 2015), interest arose in evaluating the possibilities of integrating mycorrhizal inoculants with green manures, both preceding and intercropped, and taking advantage of the benefits of both practices.

5.1. Response of *C. ensiformis* to AMF Inoculation in Different Edaphic Conditions

C. ensiformis commonly known as "jack bean" was one of crops widely used in experiments to compare the effectiveness of mycorrhizal inoculants in different soil conditions (Table 3), therefore in these new experiments the use of the AMF isolate to be inoculated was based on the pH-H_2O of the soil. The response of canavalia to inoculation was initially evaluated in nine edaphic conditions included in Table 1 (Vertic Phaozem; Eutric Cambisol; Gley Pelic Vertisol; Rhodic Lixic Eutric Nitisol; Rhodic Alumic Acrisol; Eutric Cambisol without carbonates; Dystric Cambisol and, Albic Pisoplintic Plintosol).

In all cases the inoculation produced a significant response ($p<0.01$) in the different variables evaluated and the main results are shown in Figure 6. The positive effects achieved by the inoculation on biomass production (dry matter) were remarkable, reaching average increases of 50% and always greater than 20%. It should be noted that the greatest effects were found in the soils of lower fertility and associated with the lower development of the control treatments, although the highest biomass yields in the inoculated treatments (6 - 7 t ha^{-1}) were achieved in the soils of higher fertility.

Similarly, the average increases in the amounts of macronutrients (NPK) extracted were higher than 60% and can reach after 60 days a total extraction of 170, 14 and 135 kg ha^{-1} of N, P and K respectively. In the macronutrient's concentration, the effects were also significant but more discreet, between 8 and 15% and possibly linked to the remarkable increases obtained in biomass production. The important increases in the amounts of macronutrients extracted by the inoculated canavalia must have a valuable impact on their potential as green manure.

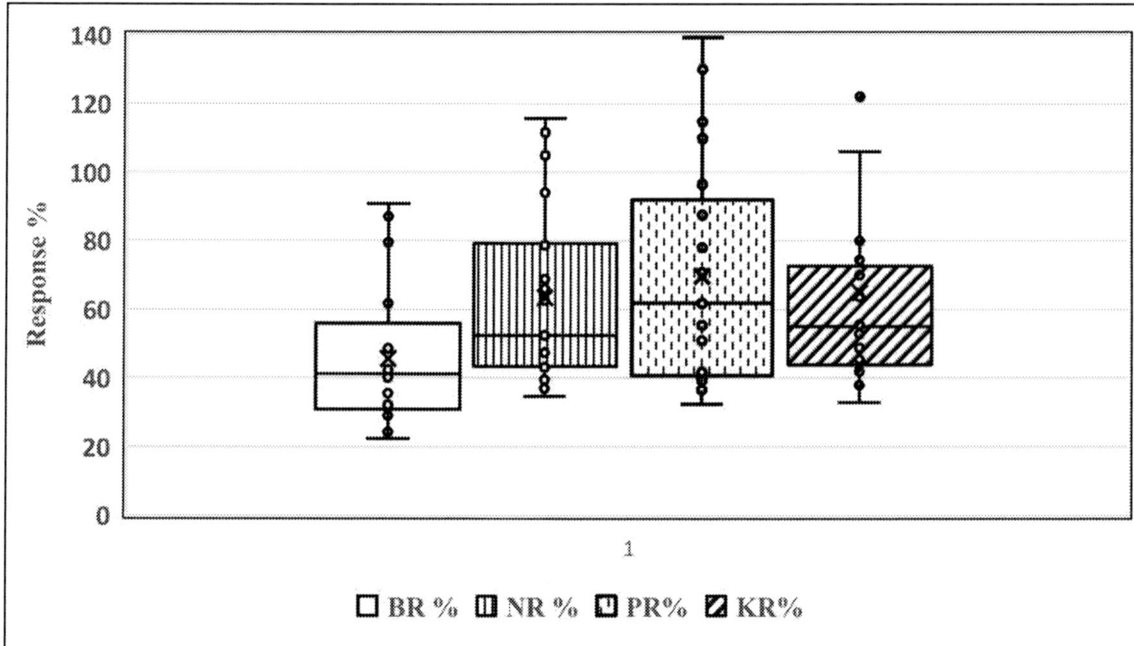

Figure based on: Bustamante et al., (2010); Martin et al., (2010, 2015); Ruiz et al., (2012); Tamayo et al., (2015); García-Rubido et al., (2017); Ojeda et al., (2018); Simó et al., (2019).

Figure 6. Main results obtained by inoculating canavalia with the efficient AMF isolate in different soil condition. Responses in aerial biomass dry matter (BR%), in the amounts of N, P and K present in it (NER%; PER%; KER%), and in the respective NPK concentrations (NCR%; PCR% and KCR%). x: Average response in each variable. N = 14. The response for each variable in each experiment was significant at least for $P < 0.01$.

Furthermore, the inoculation of the efficient isolate generally achieved average colonization percentages of around 60%, which, according to diverse authors, they are indicative of effective mycorrhizal functioning in different crops (Ruiz et al., 2012; González et al., 2015; Simo et al., 2020). Canavalia is a legume that commonly establishes an association with different species of rhizobia present in the soil. The significant contribution of effective mycorrhizal functioning to symbiotic fixation in legumes has become clear, related among others by improvements in phosphorus nutrition and other nutrients in legumes (Larimer et al., 2014; Bulgarelli et al., 2017). So the positive effects of both symbiosis are likely to be present in our results explaining the increase in biomass production, concentrations and quantities of nutrients extracted by the AMF inoculated canavalia.

However, canavalia also responds positively to inoculation with specific rhizobial isolates and the co-inoculation of effective AMF and rhizobial isolates increases the effects of single inoculations of both microorganisms (Bustamante et al., 2010; Tamayo et al., 2015), therefore, the use of co-inoculation to increase the potential of this green manure should be considered in our future work.

5.2. Benefits of the Inoculated Jack Bean with AMF Inoculant in the Nutrient Supply Schemes of Different Crops

Different experiments were developed to evaluate the effects of canavalia inoculated with efficient AMF isolates and used as a precedent and/or intercrop, on mineral fertilizer requirements, mycorrhizal performance and yields of different crops in different types of soils. The main characteristics of experiments conducted are presented in Table 10.

The canavalia inoculated with the efficient AMF isolate and used as the preceding or intercropped green manure caused positive effects in the various main crops evaluated in different types of soils (Figure 7). In all the crops, significant response ($p < 0.01$) in the yields with average of 50% were obtained in relation to the treatments that received the same medium or low dose of fertilization (F1) and that did not include inoculated canavalia. Likewise, high yields obtained were similar to those achieved when using the higher fertilization doses (100% NPK without inoculated canavalia) recommended by the Crop Technical Instructions in each crop (IIT, 2012; MINAG, 2004, 2011, 2014).

Table 10. Main characteristics of experiments on the integration of AMF-inoculated Canavalia into fertilisation schemes and technologies of different crops

Economic crop/variety (references)	Soil/AMF strain inoculated	*C. ensiformis*	Experimental period	Treatments	Fertilization 100% NPK N-P-K (kg ha^{-1})
Corn/'Francisco mejorado' (1)	Eutric Nitisol/INCAM-4	Precedent	2006-2007	Factorial arrangement 2 x 2, (canavalia inoculated or not), (corn inoculated or not).	all of the treatments received 50 kg ha^{-1} N
Tobacco/'Criollo 98' (2)	Rhodic, Acrisol/ INCAM-4	Precedent	2012-2015	Factorial arrangement 3 x 3. (50%, 75% and 100%NPK) and (canavalia-AMF, canavalia and without canavalia).	120-26-133
Banana/'FHIA-18' (3)	Eutric Cambisol/INCAM-11	Precedent and intercropped	2006-2010	Canavalia-AMF with 0, 25%, 50% and 75% FOM and other treatments as, 100% FOM, 75% FOM, 75% FOM-AMF, AMF and control without fertilizers.	100% FOM (20 kg $plant^{-1}$ of compost and 10 kg $plant^{-1}$ of sugar cane ash)
Cassava/'CMC-40' (4)	Eutric Nitisol/INCAM-4	Precedent and intercropped	2012-2013	Combinations of canavalia inoculated or not, cassava inoculated or not, all of them receiving 25% NPK and others treatments only with levels of mineral fertilizer between 0 to 100% NPK.	100 – 17 - 125
Cassava/'CMC-40'(5)	Eutric Cambisol/ INCAM-11	Intercropped	2013-2014	Factorial arrangement 2 x 2 x 2 (AMF or not in cassava), (canavalia -AMF intercropped or not), (0 25% NPK) and treatment with 100% NPK.	150 – 26 - 167

Economic crop/variety (references)	Soil/AMF strain inoculated	*C. ensiformis*	Experimental period	Treatments	Fertilization 100% NPK N-P-K (kg ha^{-1})
Mullberry/'Tigreada' (6)	Eutric Nitisol/INCAM-4	Intercropped	2008-2010	Factorial arrangement 3 x 2 (0, 50 y 100% NPK) and (canavalia-AMF and without canavalia).	200 – 44 - 83
"King-grass" (*P. purpureum*)/'Morado' (2)	Pisoplintic Plintosol/INCAM2	Precedent	2013-2014	Different combinations of canavalia inoculated or not, "king grass" inoculated or not, all of them with 70% N 50% PK and other treatments with 100% NPK and control without fertilizers.	200 – 35-125
	Eutric Nitisol/INCAM-4	Precedent	2014-2015		200 – 22-83
Corn/'Francisco mejorado' (7)	Eutric Nitisol/INCAM-4	Precedent	2007-2008	Split plot design. In the main parcels (fallow, canavalia, canavalia-AMF) in the subplots (corn inoculated or not inoculated).	without fertilizer
Tomato (8)	Eutric CambisolsINCAM-11	Precedent	2019	Factorial arrangement 2 x 2 x 2 (canavalia inoculated or no; 2 ages of transplanting and with or without AMF inoculation of tomatoes).	without fertilizer

A more complete information about each experiment can be found in the following publications: (1) Martín et al. (2009); (2) Rivera et al. (2017); (3) Simó et al. (2020); (4) Joao et al. (2017); (5) Joao et al. (2018); (6) Pentón et al. (2016); (7) Martín et al. (2012); (8) Tamayo et al. (2021).

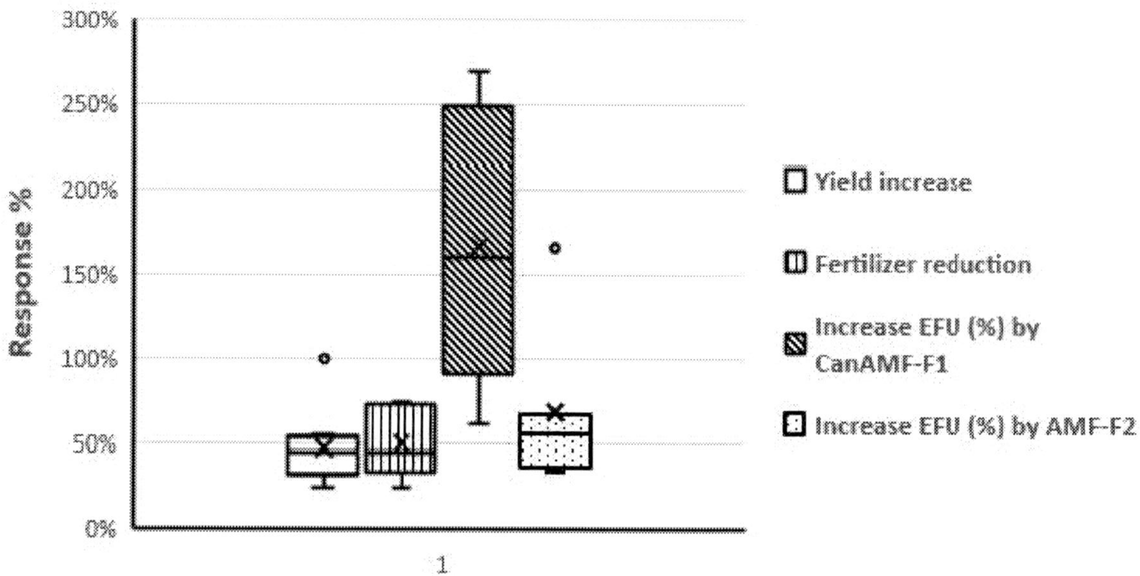

Figure elaborated from the publications referred to in Table 10.

Figure 7. Benefits of the joint management of AMF inoculants and *C. ensiformis* as green manure preceding and/or intercropped (CanAMF) in different crops in the presence of the optimal doses of fertilizer (NPK) obtained for this management (F1). For each experiment the responses in the increase in yield (%) were calculated in relation to the treatments that received F1 without CanAMF. The reduction of fertilizers (%) were calculated in relation to the 100% treatment of NPK (recommended doses for high yields by the different Crops Technical Instructive). The increase in the "Efficiency in fertilizer utilization"(EFU%) were calculated for CanAMF-F1 and for AMF-F2 (crops were AMF inoculated and F2 was the optimal fertilizer doses for this management) and both in relation to the EFU of 100% NPK treatment. The doses of 100% NPK, F2 and F1 were different for each crop, but always 100% NPK > F2> F1. x: Average response in each variable. N=8. The response in each experiment was always significant at least for P <0.01.

Therefore, in all crops a significant decrease was achieved in the amounts of mineral or organic fertilizers necessary to obtain high yields, with an average decrease of 50% and a range between 33 and 75% depending on the crop and type of soil.

It should be noted that the results found in terms of decreasing amounts of fertilizers, exceed those previously reported by the single use of AMF inoculants that ranged between 25 and 50% of the doses (Ruiz et al., 2012; Simó et al., 2015; Ruiz et al., 2016 b; Joao et al., 2017; Joao et al., 2018). Likewise, it was also found in all crops that the relationship between the estimates of "efficiency in fertilizer utilization" (EFU) and always compared to that obtained for the 100% NPK treatment were superior when AMF inoculated canavalia was used than when only AMF inoculants were applied (Figure 7). All of which supports the benefits obtained with this joint management of canavalia and AMF inoculants.

In addition, the use of inoculated canavalia allowed not only a higher yield of the crops in succession, but also a satisfactory mycorrhizal functioning in these, with average colonization percentages of around 60% (Figure 8), significantly higher ($p < 0.01$) to those obtained by these crops in succession to the non-inoculated canavalia. The satisfactory mycorrhizal functioning achieved when using the inoculated canavalia and similar to when the cultures were directly inoculated (Figure 8), demonstrate the existence of the permanence effect of the inoculant applied to the canavalia.

It also corroborates the information previously obtained (Rivera et al, 2003; Sánchez et al, 2011) that although the use of green manures increases the inoculation potential of resident mycorrhizae, this increase generally does not guarantee effective mycorrhizal function in subsequent crops. Therefore, green manure must be inoculated to enhance its benefits and effectively mycorrhize the crop in succession or associated.

No less important to support the benefits achieved by this joint management in the nutrition of the main crops, are the recent reports that in the presence of organic patches both the lengthening and the number of mycorrhizal hyphae increase, as well as that the microbiota associated with mycorrhizae intensify the decomposition rate of these organic residues (Hodge and Storer, 2015; Thyrkell et al., 2016; Bukovská et al., 2018). This favors the speed of absorption of nitrogen and other nutrients present in these residues by mycorrhizal plants.

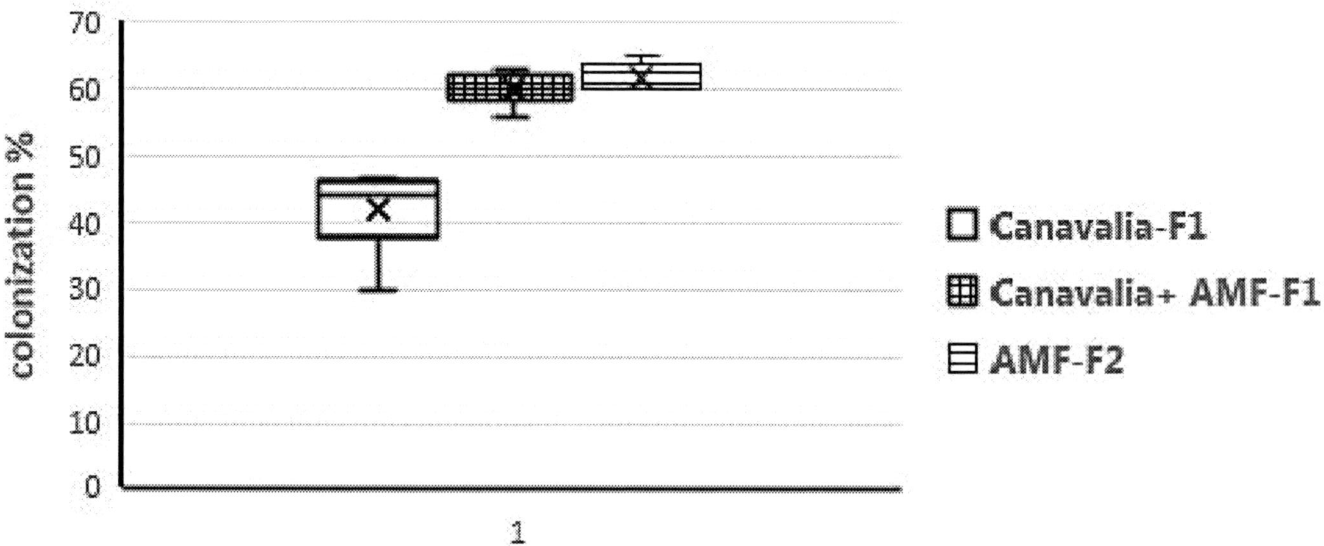

Figure elaborated from the publications referred to in Table 10.

Figure 8. Effect of joint management *C. ensiformis*: AMF on the percentages of mycorrhizal colonization of the main crops (CanAMF-F1) and compared with the percentages obtained when *C. ensiformis* is not inoculated (CanF1) and when only the main crop is inoculated with AMF and *C. ensiformis*-AMF is not used as green manure (AMF-F2). F1 and F2 optimal doses to achieve satisfactory mycorrhizal function in the presence or not of inoculated *C. ensiformis*. F2> F1. x: Average response in each variable. N= 9 (crops: corn, tobacco, banana, cassava, tomato).

Although the inoculated green manure recycles and provides significant amounts of nutrients, the effective functioning of mycorrhizae in the main crop facilitates not only the absorption of these, but also a greater absorption of nutrients from the soil and fertilizers (Yang et al., 2014; Cavagnaro et al., 2015). The above explains the decrease in the amounts of fertilizers to obtain high yields, as well as the notable increases in EFU of the main crops when inoculated canavalia is used compared to when only use of AMF inoculants.

The effect of the intercropped inoculated canavalia was also very satisfactory, guaranteeing not only high yields of the main crop, but also the fertilizer doses decreased and the EFU of the applied fertilizer was increased (Figure 7). In cassava, its employment improved the results achieved with the use of the inoculated canavalia as precedent or when the cassava was directly inoculated (Joao et al., 2017) and in all cases indicated the contribution of the green manure inoculated to the nutrition of the main crop. In the case of the mulberry in which only the intercropped canavalia was inoculated, the increases in yield, as well as the decreases in fertilizers to reach high yields and the highest values of fungal occupation reached (Figure 8) showed that it was an effective way to mycorrhizal the associated main crop. In addition quickly covers the streets, reduces the number of cleaning tasks between 70 and 80% (Pentón, 2015; Joao et al., 2018) and keeps the ground covered especially in the times of high rainfall, therefore it is also a favorable handling practice for soil conservation.

The permanence effect of the applied inoculant had been previously reported in crop sequence experiments (Rivera and Fernandez, 2003; Marrero et al., 2008) based on the generalist character of the efficient isolates with plant species (van der Heidjen et al., 2015) and with these results extend to the use of canavalia as a green manure.

This effect is important to not only simplify the handling of the inoculant and enhance the benefit of green manure, but also decrease the quantities of inoculants. For example, canavalia only requires 8 to 10 kg ha^{-1} of inoculants, much less than the 22, 26 and 120 kg ha^{-1} that require direct inoculation of cassava, banana and king-grass, respectively; and the high quantities and industriousness that also requires direct application to established crops such as mulberry and fruit plantations (20 to 60 g/plant in 3 or 4 holes around each one). In this way, the use of the permanence effect of the inoculant applied to the preceding crop allows the benefits of effective mycorrhizal performance to be extended to vegetative propagated crops and established perennials plantations, which would be prohibitive through direct inoculation of these crops (Rivera et al, 2020).

There are different documents that widely discuss the importance of green manures and aspects to take into account for their use in agricultural production systems, not only focused on family farming, but also as part of sustainable intensification technologies (Bunch, 2006; Cherr et al., 2006; Peoples et al., 2019; Royal Society, 2009); however, they do not value or warn about the importance of this joint management of green manures and mycorrhizal inoculants.

Vanlauwe et al. (2010) define the scope of integrated fertility management (ISFM) systems whose principles include the co-management of organic fertilizers and manures, the use of appropriate germplasm and integration with locally proven cultural practices. The positive effects of co-management of canavalia and mycorrhizal inoculants in the presence of low to medium doses of fertilizers (mineral or organic) on different soils and crops, ensuring high yields and increased fertilizer efficiency allow us to recommend it as a practice to be included in the ISFM approach or even in the responsible plant nutrition approach (Dobermann et al, 2022) and to move it to the stage of validation and adaptation of these results in farmers' production conditions.

6. Validation of the Results with Farmers

The working approach followed in the programme has been to validate the different results found with farmers. Rivera et al. (2020) summarise a wide group of trials and extensions carried out in about 20 crops in which different ways of inoculant application, the effectiveness of the inoculation of the efficient isolate by edaphic condition and the combination with bioproducts, green manures, organic fertilizers, among others, have been successfully validated. It is important to note that the actions with farmers, in addition to validating the proposed technologies, have a very important training component for producers. They also make it possible to enrich the proposed technologies and to confront them with different conditions and local resources, including the farmers' criteria, all of which allows the development of technologies more effective to the conditions of the farmers.

Specifically in the thematic of combined management of inoculated green manures and fertilizer requirements, the most complete work has been achieved in the production of forage for livestock, among other aspects, due to the availability of results for several years, which is essential to evaluate the sustainability of a technology.

Table 11 shows average results obtained in two entities in Villa Clara province. The forage evaluated was *Pennisetum purpureum* (king grass). This grass is propagated vegetatively and planting is done with 2 to 3 t ha^{-1} of vegetative propagules, so the direct inoculation via seed coating is a prohibitive way and therefore a previous inoculated crop is used as a way to introduce the efficient isolate of AMF (permanence effect of the applied inoculant).

Table 11. Yields of king grass 'Morado' (t ha^{-1}) by the use of *C. ensiformis* precedent inoculated with AMF. Validation assay carried out in CCS "Rubén Martínez Villena" and Agricultural Enterprise "Santo Domingo," Villa Clara

Technologies evaluated		First year (2018-2019) BDM (t ha^{-1})	Second year (2019-2020) BDM (t ha^{-1})	Third year (2020-2021 BDM (t ha^{-1})	Accumulated BDM (t ha^{-1})
1	Canavalia inoculated + 10 ha^{-1} of cattle manure	9.59	8.13	4.77	22,49
2	Canavalia without inoculation + 10 t ha^{-1} of cattle manure	7.27	6.15	2.87	16,29
3	20 ha^{-1} of cattle manure	9.72	8.93	4.96	23,61

Albic Pisoplintic Plintosol Soil pH-H2O: 4.8. isolated AMF applied *F. mosseae*/INCAM-2. BDM: king grass biomass dry matter. Four annual cuttings except in the last year when only two were given.

The nutrient source used in these entities was cattle manure. The total acreage in each validation assay was 20 ha and three technologies were compared. 1) using precedent canavalia inoculated with AMF and 10 t ha^{-1} per year of manure; 2) using precedent canavalia + 10 t ha^{-1} per year of manure (non-inoculated counterpart treatment) and 3) the technology commonly used by producers, applying 20 t ha^{-1} per year of manure. In the last year, there were difficulties in maintaining the trials due to COVID-19 consequences and, therefore, we were only able to collect data from two of the four annual cuts.

The results obtained validate that the use of inoculated canavalia as a precedent allows a 50% reduction in the amount of manure applied and maintains the productivity of the area, as well as making clear the need to inoculate the precedent canavalia in order to obtain these benefits.

Although the forage species studied respond positively to inoculation at sowing, this mycorrhizal performance does not remain effective throughout the life of the crop. In addition, the target for inoculation may be an established area of king grass. In this case it can be done in a similar way as described for the re-inoculation of established areas of *U. decumbens* or *U.* "hybrid mulato." However, it is also appropriate to use a perennial legume species that can be intercropped as an inoculated crop and thus provide benefits associated with its use as cover crop, as a way to ensure effective mycorrhizal functioning of the king grass and to improve forage composition. Although inoculated canavalia is suitable as an inoculated precedent and even re-inoculation through intercropping is effective, however as it is not a perennial legume it disappears after the second cutting and with it the benefits in mycorrhization and improved forage composition. Therefore, its effective use requires annual sowings and is therefore not recommended in this case.

Table 12 shows results of production scale trials comparing three technologies in two production units in Villa Clara province. The total area of the validation assay in each entity was 10 ha. In this case, the effectiveness of using *Teramnus labialis* inoculated as a way to apply the efficient AMF isolate in established areas of king grass 'Morado'. This is a perennial legume that is sown at a rate of 5 kg ha-^{1} of seed and intercalating one furrow of teramnus between two furrows of king grass.

Table 12. Benefits of using inoculated *Teramnus labialis* intercropping on the productivity and manure requirements of king grass 'Morado'. Validation assay carried out in CCS "Rubén Martínez Villena" and Agricultural Enterprise "Santo Domingo," Villa Clara

Technologies evaluated		First year (2018-2019)		Second year (2019-2020)		Third year (2020-2021)		Accumulated
		BDM (t ha^{-1})	% of teramnus	BDM (t ha^{-1})	% of teramnus	BDM (t ha^{-1})	% of teramnus	BDM (t ha^{-1})
1	*T. labialis* inoculated + 10 ha^{-1} of cow manure	11.87	20.5	10.75	15.3	7.39	16.0	30,01
2	*T. labialis* + 10 t ha^{-1} of cow manure	8.27	10.2	7.52	7.5	5.15	9.2	20,94
3	20 ha^{-1} of cow manure	10.12	---	9.17	---	6.58	--	25,87

Albic Pisoplintic Plintosol Soil. pH-H_2O: 4.8. isolated AMF applied *F. mosseae*/INCAM-2. BDM: king grass+teramnus biomass dry matter.% of *T. labialis*. Refers to the percentage of the area occupied by this species. Four annual cuttings except in the last year when only two were given.

The results were very favorable, the proposed technology using inoculated teramnus resulted in biomass yields even higher than those commonly used by producers (+16%) and decreasing the annual amounts of manure by 50%. As manure is not enough to apply in all king grass areas of the country, this result allows doubling the areas with high yields and therefore doubling the efficiency of fertilization.

The trials also establish the need to inoculate the teramnus to achieve the yield benefits and manure savings. Inoculation improves the survival and occupancy of teramnus in the plantation and after three years effects are maintained and it has not yet been necessary to re-inoculate. The management of teramnus was also carried out by cuts that were made to coincide with those of king grass to simplify management and improve the forage composition.

Conclusion

A positive and reproducible response to "simple" inoculation of AMF isolates was found in different crops and soil types, with differentiated effects among them. The efficiency of these inoculants depended on the pH-H_2O of the soil or substrate where the mycorrhizal roots are located. In the pH range that conditions the efficiency of the inoculant, all crops dependent on arbuscular mycorrhization establish an efficient symbiosis, although the degree of response will depend on the crop. Compatibilities between inoculants-crops that are presented do not originate responses superior to those obtained when inoculating the efficient isolate based on the pH-H_2O of the soil, therefore the criteria of using pH as a basis for the recommendation of simple inoculants to crops do not vary. The differentiated management of these inoculants allows a successful application to crops in soils that are in the pH-H_2O range of 4.7 to 8.1.

The "efficient" isolate inoculation allows reducing the amounts of fertilizers (NPK) or organic fertilizers required by the crops to obtain high yields and satisfactory nutritional status; however, for the inoculation to originate a satisfactory mycorrhizal performance and be effective, it generally requires low or medium doses of fertilizers, doses lower than these limit the mycorrhizal performance and high doses inhibit mycorrhization, therefore the effectiveness of the inoculant depends on nutrient supply. The amount of fertilizer reduced depends on the crop, the nutrient in question and its availability in the soil.

	Canavalia inoculated with INCAM-11 and intercropped in a field of cassava cultivar CMC-40 on a farmer's farm. CCS Niceto Pérez, Mayabeque. Eutric Cambisol soil with pH-H2O > 7.5. In this case an increase in cassava yield of 10 t ha-1 (+ 50%) was found and the weeds cleanings were eliminated. *Photo and unpublished data courtesy of MSc. Anicel Delgado Álvarez.*
	Canavalia intercropped and inoculated with INCAM-11 in banana experiment FHIA-18 on Eutric Cambisol soil in INIVIT experimental areas. *Photo courtesy of Dr. J. Simó González.*
	Use of teramnus inoculated with INCAM-2 and intercropped in area of *P. purpureum* cv. Common. Albic Pisoplintic Plintosol soil with pH-H2O of 4.7. Experimental Area of Forage and Pasture Station/Villa Clara. *Photo courtesy of MSc. Juan Ramírez Pedroso.*

Figure 9. (Continued).

Plantation of *C. canephora* with canavalia intercropped in III Frente, Stgo de Cuba. Note the high biomass production which, in addition to preventing undesirable vegetation, is a good cover, recycles and provides nutrients and when inoculated maintains effective mycorrhizal functioning of the coffee plant. *Photo courtesy of Dr. Carlos Bustamante González.*

Figure 9. Examples of the feasibility of the use of AMF-inoculated Canavalia in agricultural production.

The joint management of mycorrhizal inoculants and *C. ensiformis* used as inoculated green manure, either precedent and/or intercropped not only enhances the benefits obtained individually by the use of one or another "practice" and related among others with decreases in the quantities of the fertilizer requirements to guarantee an adequate nutritional status and satisfactory yields. It also becomes an effective way to introduce the AMF inoculants in the agroecosystems, guaranteeing an effective mycorrhizal functioning of the mycotrophic associate cultures, with greater impact in those that require such high amounts of mycorrhizal inoculants that are not economically feasible and also to ensure an effective inoculation in established plantations. This joint management is valid when are inoculated isolates of AMF that have a generalist character with the crops and using them in the edaphic conditions in which they are effective.

Validation results on a productive scale corroborate the experimental results obtained, allowing us to recommend their use at least in the forage areas. The information obtained is an example of the using feasibility of mycorrhizal inoculants in agricultural production technologies.

Acknowledgments

To Dr Pedro Sanchez for his interest and motivation to prepare this summary of the research carried out. To Dr. Ricardo Herrera† pioneer of mycorrhizal research in Cuba. To the researchers and specialists who participate in the Cuban network for the management of arbuscular mycorrhizal symbiosis in agroecosystems.This investigation was financed for the Fund for Science and Innovation (FONCI) of Cuba.

References

Arora NK, Mehnaz S, Balestrini R (2016) *Bioformulations: for Sustainable Agriculture.* Springer, New Delhi, India, 1-299. doi: 10.1007/978-81-322-2779-3.

Asghari HR, Cavagnaro TR (2014) Mycorrhizas effects on nutrient interception in two riparian grass species. *Eurasian Journal of Soil Science,* 3:274-285. doi:10.18393/ejss.10069.

Azcon R, Ambrosano E, Charest C (2003) Nutrient acquisition in mycorrhizal lettuce plants under different phosphorus and nitrogen concentration. *Plant Science,* 165:1137-1145. doi:10.1016/S0168-9452(03)00322-4.

Bagyaraj DJ, Sharma MP, Maiti D (2015) Phosphorus nutrition of crops through arbuscular mycorrhizal fungi. *Current Science,* 108:1288-1293. https://www.jstor.org/stable/24905490.

Bever JD, Morton JB, Antonovics J, Schultz PA (1996) Host-dependent sporulation and species diversity of arbuscular mycorrhizal fungi in a mown grassland. *Journal of Ecology,* 84(1):71-82. doi:10.2307/2261701.

Bhandari P, Garg N (2017) Dynamics of arbuscular mycorrhizal symbiosis and its role in nutrient acquisition: an overview. In: Varma A, Prasad R, Tuteja N (eds.) *Mycorrhiza - Nutrient Uptake, Biocontrol, Ecorestoration.* Springer International Publishing AG, pp. 21–43. doi:10.1007/978-3-319-68867-1_2.

Bukovská P, Bonkowski M, Konvalinková T, Beskid O, Hujslová M, Püschel D, Řezáčová V, Gutiérrez-Núñez MS, Gryndle M, Jansa J (2018) Utilization of organic nitrogen by arbuscular mycorrhizal fungi —is there a specific role for protists and ammonia oxidizers? *Mycorrhiza,* 28:269-283. doi:10.1007/s00572-018-0825-0.

Bulgarelli RG, Castro MF, Ribeiro RV, López de Andrade S (2017) Mycorrhizae enhance nitrogen fixation and photosynthesis in phosphorus starved soybean (*Glycine max* L. Merrill). *Environmental and Experimental Botany,* 140:26-33. doi:10.1016/j.envexpbot.2017.05.015.

Bunch R (2006) Green manures/cover crops for recuperating soils and maintaining soil fertility in the tropics. In: Uphoff N, Ball AS, Fernandes E, Herren H, Husson O, Laing M, Palm C, Pretty J, Sanchez P, Sanginga N, Thies J (Eds.) *Biological Approaches to Sustainable Soil Systems.* CRC Press, Taylor& Francis Group, Florida, USA, pp. 439-452.

Bustamante C, Rivera R, Pérez G, Viñals R (2010) Co-inoculation of rhizobium strains and AM fungi in Cambisol soil promotes the growth of *Canavalia ensiformis. Café y Cacao*, 9(2):5-9.

Cabrera A. (1992) *Agrochemical Characteristics of Ferralitic Soils for Sugarcane Planting in Cuba.* Doctoral Thesis in Agricultural Science. National Sugarcane Research Institute, INICA, Ministry of Sugar, Cuba, pp. 1-100.

Cavagnaro TR, Bender SF, Asghari HR, van der Heijden MGA (2015) The role of arbuscular mycorrhizas in reducing soil nutrient loss. *Trends in Plant Science,* 20:283–290. doi:10.1016/j.tplants.2015.03.004.

Cherr CM, Scholberg JMS, McSorley R (2006) Green manure approaches to crop production: asynthesis. *Agronomy Journal,* 98:302–319. doi:10.2134/agronj2005.0035.

Chu Q, Wang X, Yang Y, Chen F, Zhang F, Feng G (2013) Mycorrhizal responsiveness of maize (Zea mays L.) genotypes as related to releasing date and available P content in soil. *Mycorrhiza,* 23:497-505. doi: 10.1007/s00572-013-0492-0.

Crews TE, Peoples MB (2004) Legume versus fertilizer sources of nitrogen:ecological trade-offs and human needs. *Agriculture, Ecosystems and Environment,* 102:279-297. doi:10.1016/j.agee.2003.09.018.

Dobermann A, Bruulsema T, Cakmak I, Gerard B, Majumdar K, McLaughlin M, Reidsma P, Vanlauwe B, Wollenberg L, Zhang F, Zhang X (2022) Responsible plant nutrition: A new paradigm to support food system transformation. *Global Food Security,* 33:100636. doi: 10.1016/j.gfs.2022. 100636.

Dodd JC, Thompson BD (1994) The screening and selection of inoculant arbuscular mycorrhizal and ectomycorrhizal fungi. *Plant and Soil,* 159:149-158. doi:10.1007/BF00000104.

Espinosa A, Ruiz L, Rivera R, Espinosa E (2015) Nitrogen and arbuscular mycorrhizal fungi (AMF) effect on two commercial sweet potato clones on Eutric Cambisol soil. *Centro Agrícola,* 42(2):39-46.

Espinosa A, Ruiz L, Rivera R, Espinosa E (2016) Inoculation of HMA and phosphoric fertilizer requirements in *Ipomea batata* (L.) Lam. *Centro Agrícola,*43 (3):37-45.

Espinosa A, Ruiz L, Rivera R, Espinosa E (2017) Effect of potassium doses and arbuscular micorrizing fungi in two sweet batata comercial clones on Eutric Cambisol soil. *Agrotecnia de Cuba,* 41(1):1-16.

Falcón E, Riera M, Rodríguez O (2013) Effect of the inoculation of mycorrhizal fungi on forest production postures in two types of soils. *Cultivos Tropicales,* 34(3):32-39.

Fernandes de Sousa F, Lopez do Carmo D, Souza Carneiro JE, Urquiaga S, Silva Santos RH (2016) Legumes as green manure for common bean cultivated in two growing seasons at southeast Brazil. *African Journal of Agricultural Research,* 11: 4953-4958. doi:10.5897/AJAR2016.11689.

Fernández F (1999) *Management of Mycorrhizal Associations in the Production of Coffee Seedlings in Some Soils of Cuba.* Doctoral thesis in Agricultural Science. Havana Agricultural University, Inca, pp. 1-102.

Fernández F, Gómez R, Vanegas LF, Martínez MA, de la Noval B, Rivera R (2000) Mycorrhizal Inoculation Products. Certificado Nro. 22641. Oficina Cubana de la Propiedad Industrial, La Habana.

Fundora LR, Rivera R, Martín JV, Calderón A, Torres A (2011) Efficient use of strains arbuscular mycorrhizal fungi in the development of avocado rootstocks in a substrate made up of Ferrallitic soil and filter-cake. *Cultivos Tropicales,* 32(2):23-29.

Furrazola E, Torres-Arias Y, Ojeda-Quintana L, Fors RO, Rodríguez-Rodríguez R, Ley-Rivas JF, Mena A, González-González S, Berbara RLL, Queiroz MB, Hamel C, Goto BT (2021) Research on arbuscular mycorrhizae in Cuba: a historical review and future perspectives. *Studies in Fungi,* 6(1): 240–262. doi 10.5943/sif/6/1/16.

García M, Rivera R, Cruz Y, Acosta Y, Cabrera J (2017) Response of *Canavalia ensiformis* (L.) to inoculation with different arbuscular mycorrhizae fungi in a lixiviated Ferralitic soil. *Cultivos Tropicales,*38(1):16-21.

García M, Treto E, Álvarez M (2000) Green fertilizer: an alternative to nitrogen economy in potato cultivation. The interaction of green fertilizer and nitrogen dosage. *Cultivos Tropicales,* 21(1):13-19.

García M, Treto E, Álvarez M (2002) Comparative study on different species of green manures and their effects on maize cultivation. *Cultivos Tropicales,* 22(3):19-30.

George E (2000) Nutrient uptake. In: Kapulnik Y, Douds DD (Eds.) *Arbuscular Mycorrhizas: Physiology and Function.*Springer, Dordrecht. pp:307-343. doi:10.1007/978-94-017-0776-3_14.

Gerdemann JW, Nicolson TH (1963) Spores of mycorrhizal Endogone species extracted from soil by wet sieving and decanting. *Transactions of the British Mycological Society,* 46 (2): 235–44. doi: 10.1016/S0007-1536(63)80079-0.

González PJ (2014) *Effective Management of Arbuscular Mycorrhizal Symbiosis by Inoculating of Brachiaria Forage and Mineral Fertilization.* PhD thesis. Havana Agricultural University, pp. 1-90.

González PJ, Ramírez JE, Rivera R, Hernández -Jiménez A, Crespo G (2018) Critical levels of phosphorus in the soil for forage legumes, inoculated with arbuscular mycorrhizal fungi. *Cuban Journal of Agricultural Science,* 50 (2):315-320.

González PJ, Ramírez JF, Morgan O, Rivera R, Plana R (2015b) Contribution of arbuscular mycorrhizal inoculation to the reduction of phosphorus fertilization in *Brachiaria decumbens. Cultivos Tropicales,*36(1):135-142.

González PJ, Ramirez JF, Rivera R, Hernández A, Plana R, Crespo G (2015a) Management of arbuscular mycorrhizal inoculation for the establishment, maintenance and recovery of grasslands. *Cuban Journal of Agricultural Science,* 49(4): 535–540.

González PJ, Ramírez JF, Rivera R, Hernández-Jiménez A, Crespo G (2016). Effectiveness of arbuscular mycorrhizal fungi inoculation on two leguminous forages planted in two soils. *Tropical Grasslands-Forrajes Tropicales,*4(2):82-90.

Guo J, Ling N, Chen Z, Xue C, Li L, Liu L, Gao L, Wang M, Ruan Y, Guo S, Vandenkoornhuyse P, Shen Q (2020) Soil fungal assemblage complexity is dependent on soil fertility and dominated by deterministic processes. *New Phytologist,*226(1): 232-243. doi: 10.1111/nph.16345.

Habte M. (2006). The roles of arbuscular mycorrhizas in plant and soil health. In: Uphoff N, Ball AS, Fernandes E, Herren H, Husson O, Laing M, Palm C, Pretty J, Sanchez P, Sanginga N, Thies J (Eds.) *Biological Approaches to Sustainable Soil Systems.* CRC Press, Taylor& Francis Group, Florida, USA. pp.129-147.

Hamel C, Dalpé Y, Furlan V, Parent S (1997) Indigenous populations of arbuscular mycorrhizal fungi and soil aggregate stability are major determinants of leek (*Allium porrum* L.) response to inoculation with *Glomus intraradices* Schenck & Smith or

Glomus versiforme (Karsten) Berch. *Mycorrhiza,*7:187–196. doi: 10.1007/s005720 050180.

Hamel C, Strullu DG (2006) Arbuscular mycorrhizal fungi in field crop production: Potential and new direction. *Canadian Journal of Plant Science,*86: 941–950. doi:10.4141/P05-099.

Hayman DS, Tavares M (1985) Plant growth responses to vesicular-arbuscular mycorrhiza. XV. Influence of soil pH on the symbiotic efficiency of different endophytes. *New Phytologist,* 100 (3):367-377. doi: 10.1111/j.1469-8137.1985.tb02786.x.

Helgason T, Fitter AH (2009) Natural selection and the evolutionary ecology of the arbuscular mycorrhizal fungi (Phylum Glomeromycota). *Journal of Experimental Botany,* 60:2465-2480.doi:10.1093/jxb/erp144.

Helgason T, Merryweather JW, Young JPW, Fitter AH (2007) Specificity and resilience in the arbuscular mycorrhizal fungi of a natural woodland community. *Journal of Ecology,* 95:623-630. doi:10.1111/j.1365-2745.2007.01239.x.

Hernández-Jiménez A, Pérez JM, Bosch D, Castro N (2015) *Cuban soil classification.* Ediciones INCA, Mayabeque, Cuba, pp, 1-93.

Hodge A, Storer K (2015) Arbuscular mycorrhiza and nitrogen: implications for individual plants through to ecosystems. *Plant and Soil,* 386:1-19. doi: 10.1007/s11104-014-2162-1.

Hoeksema JD, Chaudhary VB, Gehring CA, Collins N, Karst J, Koide RT, Pringle A, Zabinski C, Bever JD, Moore JC, Wilson GWT, Klironomos JN, Umbanhowar J (2010) A meta-analysis of context-dependency in plant response to inoculation with mycorrhizal fungi. *Ecology Letters,* 13:394–407. doi: 10.1111/j.1461-0248.2009.014 30.x.

Iturralde-Vinent MA, Garcia-Casco A (2012) *Geological Compendium of Cuba and the Caribbean.* Science and Technology Press, Library Research Institute, second edition, pp. 1-150.

Janos DP (2007) Plant responsiveness to mycorrhizas differs from dependence upon mycorrhizas. *Mycorrhiza,* 17:75-91. doi:10.1007/s00572-006-0094-1.

Jansa J, Forczek ST, Rozmoš M, Püschel D, Bukovská P, Hršelová H (2019) Arbuscular mycorrhiza and soil organic nitrogen: network of players and interactions.*Chemical and Biological Technologies in Agriculture,* 6:10.doi:10.1186/s40538-019-0147-2.

João JP, Espinosa A, Ruiz L, Simó J, Rivera R (2016) Effectiveness of AMF strains on cassava crop (*Manihot esculenta* Crantz) in two types of soils. *Cultivos Tropicales,* 37(1):48-56.

Joao JP, Rivera R, Martín G, Riera M, Simó JE (2017) Integrated nutrition system with AMF, green manure and mineral fertilizers in *Manihot esculenta* Crantz. *Cultivos Tropicales,* 38 (3):117-128.

Joao JP, Rivera R, Martin, GM (2018) Mycorrhizal inoculation, *Canavalia ensiformis* and fertilization in cassava cultivation. Spanish Academic Press, VDM Publishing, Alemania, pp. 1-157.

Johnson NC, Graham JH, Smith FA (1997) Functioning of mycorrhizal associations along the mutualism–parasitism continuum. *New Phytologist,* 135:575–585. doi:10.1046/j. 1469-8137.1997.00729.x.

Kang BT, Islam R, Sanders FE, Ayanaba A (1980) Effect of phosphate fertilization and inoculation with VA-mycorrhizal fungi on performance of cassava (*Manihot esculenta* Crantz) grown on an Alfisol. *Field Crops Research,* 3:83-94. https://doi.org/10.1016/0378-4290(80)90009-X.

Lambers H; Raven J; Shaver GA, Smith SE (2008) Plant nutrient-acquisition strategies change with soil age. *Trends in Ecology & Evolution,* 32:95-103. doi:10.1016/j.tree.2007.10.008.

Larimer AL, Clay K, Bever JD (2014) Synergism and context dependency of interactions between arbuscular mycorrhizal fungi and rhizobia with a prairie legume. *Ecology,* 95:1045–1054. doi: 10.1890/13-0025.1.

Le Mire G, Nguyen ML, Fassotte B, du Jardin P, Verheggen F, Delaplace P, Jijakli MH (2016) Review: Implementing plant biostimulants and biocontrol strategies in the agroecological management of cultivated ecosystems. *Biotechnologie, Agronomie, Societe et Environment,* 20(S1):299-313. doi: 10.25518/1780-4507.12717.

Lehmann A, Leifheit EF, Rillig MC (2017) Mycorrhizas and soil aggregation. In: Johnson NC, Gehring C, Jansa J (Eds.) *Mycorrhizal Mediation of Soil.* Elsevier, pp. 241-262. doi:10.1016/B978-0-12-804312-7.00014-0.

Lekberg Y, Koide RT (2005) Is plant performance limited by abundance of arbuscular mycorrhizal fungi? A meta-analysis of studies published between 1988 and 2003. *New Phytologist,* 168: 189-204.doi: 10.1111/j.1469-8137.2005.01490.x.

Lesueur D, Deaker R, Herrmann L, Brau L, Jansa J (2016). The production and potential of biofertilizers to improve crop yields. In: Arora N, Mehnaz S, Balestrini R (Eds.) *Bioformulations for Sustainable Agriculture.* Springer India, pp. 71-92. doi: 10.1007/978-81-322-2779-3_4.

Liu A, Hamel C, Elmi A, Costa C, Ma BL, Smith DL (2002) Concentrations of K, Ca and Mg in maize colonized by arbuscular mycorrhizal fungi under field conditions. *Canadian Journal of Soil Science,* 82:271-278. doi: 10.4141/S01-022.

Liu A, Hamel C, Hamilton RI, Ma BL, Smith DL (2000) Acquisition of Cu, Zn, Mn and Fe by mycorrhizal maize (*Zea mays* L.) grown in soil at different P and micronutrient levels. *Mycorrhiza,* 9:331-336. doi: 10.1007/s005720050277.

Lópes ES, de Toledo SV, Hiroce R, Dias R, Oliveira ED (1983) Effects of the mycorrhizal fungus *Gigaspora margarita* on the development of coffee seedlings c.v. Mundo Nova in field conditions. In: 10 *Congreso Brasileiro de Pesquisas Cafeeiras*. Poços de Caldas, Anais, Río de Janeiro, IBC/GERCA,pp. 122-123.

Maiti D, Toppo NN (2017) Capturing plant genetic potential of upland rice for exploiting arbuscular mycorrhiza responsiveness to improve rice variety for higher phosphorus (P) acquisition under P limiting environments. In: Varma A, Prasad R, Tuteja N (Eds.) *Mycorrhiza - Nutrient Uptake, Biocontrol, Ecorestoration.* Springer, Cham., pp. 45-73. doi:10.1007/978-3-319-68867-1_3.

Marrero Y, Simó J, Ruiz L, Rivera R, Plana R (2008) Effects of tillage on effective mycorrhizal symbiosis management in crop sequences in Eutric Cambisol soil. *Cultivos Tropicales,* 29(2): 11-15.

Marschner H, Dell B (1994) Nutrient uptake in mycorrhizal symbiosis. *Plant and Soil* 159:89-102.doi:10.1007/BF00000098.

Martín GM, Arias L, Rivera R (2010) Selection of more effective AMF isolates for *Canavalia ensiformis* growing on a Red Ferralitic soil. *Cultivos Tropicales,* 31(1):27-31.

Martín GM, Reyes R, Ramírez JF (2015) Coinoculation of *Canavalia ensiformis* with Rhizobium and arbuscular mycorrhizal isolates in two soils from Cuba. *Cultivos Tropicales,* 36(2):22–29.

Martín GM, Rivera R, Arias L, Rentería M (2009) Effects of *Canavalia ensiformis* and arbuscular mycorrhiza on maize cultivation. *Revista Cubana de Ciencia Agrícola,* 43(2):191-195.

Martín GM, Rivera R, Pérez A, Arias L (2012) *Canavalia ensiformis* response to mycorrhizal inoculation with *Glomus cubense* (isolate INCAM-4), their permanency effect in the maize crop. *Cultivos Tropicales*, 33(2):20-28.

Mensah JA, Koch AM, Antunes PM, Kiers T, Hart M (2015) High functional diversity within species of arbuscular mycorrhizal fungi is associated with differences in phosphate and nitrogen uptake and fungal phosphate metabolism. *Mycorrhiza,* 25(7):533-546.doi:10.1007/s00572-015-0631-x.

MINAG (1987) *Technical Instruction Manuals of Coffee and Cocoa crops.* Direction of Coffee and Cocoa of the Ministry of Agriculture. Havana, Cuba, pp. 1-208.

MINAG (2004) *Technical Instruction Manuals of Tropical Roots and Tubers*. Instituto de Investigaciones de Viandas Tropicales (INIVIT). Havana, Cuba, pp.1-50.

MINAG (2011) *Guidance on Banana Cultivation Techniques.* Instituto de Investigaciones de Viandas Tropicales (INIVIT). First edition. Havana, Cuba, pp. 1-12.

MINAG (2012) *Technical Guidance for Tropical Roots and Tubers.* Havana. Ministry of Agriculture, pp. 1-120.

MINAG (2014) *Animal Husbandry Technical Manual.* Havana. CIMA edition: second edition. Pp. 1-106.

Morejón M, Herrera JA, Ayra C, González PJ, Rivera R, Fernández Y, Peña E, Téllez P, Rodríguez-de la Noval C, de la Noval-Pons B (2017) Alternatives in nutrition of transgenic maize FR-Bt1 (*Zea mays* L.): response in growth, development and production. *Cultivos Tropicales,* 38(4): 146-155.

Mosse B (1973) Advances in the study of vesicular-arbuscular mycorrhiza. *Annual Review Phytopathology,*11: 171-196.doi:10.1146/annurev. py.11.090173.001131.

Muñoz S (1997) *Chemical and physicochemical properties of Cambisol soil planted with sugarcane and their relationship with sugarcane yiel*d. Master's thesis. Havana Agricultural University, Cuba, pp. 1-87.

Oehl F, Laczko E, Bogenrieder A, Stahr K, Bösch R, van der Heijden M, Sieverding E (2010) Soil type and land use intensity determine the composition of arbuscular mycorrhizal fungal communities. *Soil Biology and Biochemistry,* 42:724-738.

Ojeda L, González PJ, Rivera R, Furrazola E, de la Rosa JJ, Rodríguez YI, González L, Hernández ME (2018) Inoculation with arbuscular mycorrhizal fungi species on *Canavalia ensiformis* (L.) D.C in the stage of establishing feed bank. *Pastos y Forrajes,*41(3):182-188.

Opik M, Mora M (2012) Missing nodes and links in mycorrhizal networks. *New Phytologist,*194(2):304-306. doi: 10.1111/j.1469-8137.2012.04121.x.

Ortas I (2012) The effect of mycorrhizal fungal inoculation on plant yield, nutrient uptake and inoculation effectiveness under long-term field conditions. *Field Crop Research,* 125:35–48. doi:10.1016/j.fcr.2011.08.005.

Ortas I, Rafique M (2017) The mechanisms of nutrient uptake by arbuscular mycorrhizae. In: Varma A, Prasad R, Tuteja N (Eds.) *Mycorrhiza - Nutrient Uptake, Biocontrol, Ecorestoration.* Springer International Publishing AG, pp. 1-20. doi:10.1007/978-3-319-68867-1_3.

Ouzounidou G, Skiada V, Papadopoulou KK, Stamatis N, Kavvadias V, Eleftheriadis E, Gaitis F (2015) Effects of soil pH and arbuscular mycorrhiza (AM) inoculation on growth and chemical composition of chia (*Salvia hispanica* L.) leaves. *Brazilian Journal of Botany,* 38:487-495. doi: 0.1007/s40415-015-0166-6.

Paneque VM, Calaña JM, Calderón M, Borges Y, Hernández TC, Caruncho CM (2010) *Technical Manual for Soil, Leaf, Organic Fertilizer and Fertilizer Analysis.* Ed. Ediciones INCA La Habana, Cuba, PP. 1-157.

Parihar M, Rakshit A, Meena VS, Gupta VK, Rana K, Choudhary M, Tiwari G, Mishra PK, Pattanayak A, Bisht JK, Jatav SS, Khati P, Jatav HD (2020) The potential of arbuscular mycorrhizal fungi in C cycling: A review. *Archives of Microbiology,* 202:1581-1596. doi:10.1007/s00203-020-01915-x.

Pellegrino E, Opik M, Bonari E, Ercoli L (2015) Responses of wheat to arbuscular mycorrhizal fungi: A meta-analysis of field studies from 1975 to 2013. *Soil Biology & Biochemistry,* 84:210-217. doi: 10.1016/j.soilbio.2015.02.020.

Pellegrino E, Turrini A, Gamper HA, Cafà G, Bonari E, Young JPW, Giovannetti M (2012) Establishment, persistence and effectiveness of arbuscular mycorrhizal fungal inoculants in the field revealed using molecular genetic tracing and measurement of yield components. *New Phytologist,* 194(3):810–822. doi:10.1111/j.1469-8137.2012.04090.x

Pentón-Fernández G (2015) *The effect of Canavalia ensiformis (L.) inoculated with arbuscular mycorrhizal fungi and mixed with mineral fertilizers on mulberry [Morus alba (L.)] feed production*. Phd Thesis. Havana Agricultural University, pp. 1-99.

Pentón-Fernández G, Martín-Martín GJ, Rivera R (2016) Effect of the spatial arrangement and the intercropping with mycorrhiza-inoculated *Canavalia ensiformis* on the agroproductive response of *Morus alba. Pastos y Forrajes*, 39(3):92-99.

Peoples MB, Hauggaard-Nielsen H, Huguenin-Elie O, Jensen ES, Justes E, Williams M (2019) The contributions of legumes to reducing the environmental risk of agricultural production. In: Lemaire G, Carvalho PCDF, Kronberg S, Recous S (Eds.). *Agro-ecosystem Diversity: Reconciling Contemporary Agriculture and Environmental Quality.* Elsevier, Academic Press, pp. 123–143.

Plana R, González PJ, Rivera R, Varela M, Álvarez MA (2016) Forage production based on triticale (*X. triticosecale* Wittmack) in Lixic Ferralic Nitisol with varying nitrogen doses and arbuscular mycorrhizal fungal inoculation. *Cultivos Tropicales,* 37(2):22-32.

Pozo MJ, Verhage A, García-Andrade J, García J, Azcón-Aguilar C (2009) Priming plant defence against pathogens by arbuscular mycorrhizal fungi. In: Azcón-Aguilar C, Barea JM, Gianinazzi S, Gianinazzi-Pearson V (Eds.) *Mycorrhizas-Functional*

Processes and Ecological Impact. Springer-Verlag Berlin Heidelberg, pp. 123-136. doi:10.1007/978-3-540-87978-7.

Ramírez J, González PJ, Salazar X, Llanes D, Rivera R, Hernández-Jiménez A, Plana R (2017) Arbuscular–mycorrizal inoculation and reduction of organic and nitrogen fertilization in *Megathyrsus maximus* cv. Likoni. *Pastos y Forrajes,* 40(2):108-117.

Ramírez J, González PJ, Salazar X, Rivera R, Hernández-Jiménez A, Plana R (2016) Effects of lime treatment and HMA inoculation on yield of *Brachiaria* hibrido Bro1752 cv. Yacaré. *XX Congreso Cientifico Internacional de Instituto Nacional de Ciencias Agricolas, Nov 23 al 25 del 2016.*

Rillig MC, Aguilar-Trigueros CA, Camenzind T, Cavagnaro RT, Degrune F, Hohmann P, Lammel DR, Mansour I, Roy J, van der Heijden MGA, Yang G (2019). Why farmers should manage the arbuscular mycorrhizal symbiosis?. *New Phytologist,* 1-5. doi: 10.1111/nph.15602.

Rivera R, Fernández F (2006) Inoculation and management of mycorrhizal fungi within tropical agroecosystems. In: Uphoff N, Ball AS, Fernandes E, Herren H, Husson O, Laing M, Palm C, Pretty J, Sanchez P, Sanginga N, Thies J (Eds.) *Biological Approaches to Sustainable Soil Systems.* CRC Press, Taylor& Francis Group, Boca Raton, Florida, USA, pp. 479-489.

Rivera R, Fernández F, Fernández K, Ruiz L, Sánchez C, Riera M (2007) Advances in the management of effective arbuscular mycorrhizal symbiosis in tropical ecosystesm. In: Hamel C, Plenchette C(Eds.) *Mycorrhizae in Crop Production.* Haworth Press, Binghamton, NY. pp. 151-196.

Rivera R, Fernández F, Hernández-Jiménez A, Martín JR, Fernández K (2003). Effective management of mycorrhizal symbiosis, a pathway to sustainable agriculture. *Case Study: The Caribbean. Ediciones INCA*, La Habana, Cuba, pp. 1-166.

Rivera R, Fernández-Martín F. Ruiz L, González- Cañizares PJ, Rodríguez-YonY, Pérez-Ortega E, et al. (2020). *Management, Integration and Benefits of EcoMic® Mycorrhizal Biofertlizer in Agricultural Production.* Ediciones INCA, Mayabeque, Cuba, 155 p. https://www.researchgate.net/publication/340223155

Rivera R, Simó J, Martín G, García M, Espinosa A, Nápoles MC, Hernández I, Ruíz M, Ruiz L, Pérez E, Rodríguez Y, Ramírez J, González PJ (2017) The final report of the project "Joint management and impact of mycorrhizal biofertilizers and other biological products on agricultural production of different crop. Code: P131L H0010003. *Instituto Nacional de Ciencias Agrícolas*, 135 p. http://dx.doi.org/10.131 40/RG.2.2.14926.48967

Rodríguez A, Sanders IR (2015) The role of community and population ecology in applying mycorrhizal fungi for improved food security. *The International Society for Microbial Ecology Journal,* 9: 1053-1061. doi:10.1038/ismej.2014.207.

Rodríguez Y, Dalpé Y, Séguin S, Fernández K, Rivera R (2011) *Glomus cubense sp.nov.,* an arbuscular mycorrhizal fungus from Cuba. *Mycotaxon,* 118:337-347. doi:10.524 8/118.337.

Rodriguez Y, Maistro-Patreze C, Saggin-Junior OJ, Rivera R, Quiñones M, Haesart G, van Tuinen D (2021) Development of a taxon-discriminating molecular marker to trace and quantify a mycorrhizal inoculum in roots and soils of agroecosystems. *Folia Microbiologica,* 66:371-384. doi:10.1007/s12223-020-00844-y.

Rosales PR, González PJ, Ramírez JF, Arzola J (2017) Selectioni of arbuscular mycorrhizal fungi efficient strains for guinea grass (*Megathyrsus maximus* cv. likoni). *Cultivos Tropicales,* 38(1): 24-30.

Rosolem CA, Vicentini JP, Steiner F (2012) Potassium supply as affected by residual potassium fertilization in a Cerrado Oxisol. *Revista Brasileira de Ciencia do Solo,* 36:1507-1515. doi:10.1590/S0100-06832012000500015.

Ruiz L, Armario D, Rivera R, Espinosa A, Simó J, Espinosa E (2016b) Effect of nitrogen, phosphorus and potassium on banana cultivation. *Rev Agricultura Tropical,*2(1):1-8.

Ruiz L, Espinosa A, Camejo M, Simó JE, Rivera R (2016a) Effects of N, P, K, and mycorrhizal application on cassava growth in carbonate brown soil. *Rev Agricultura Tropical,* 2(2):1-13.

Ruiz L, Simó J, Rodríguez S, Rivera R (2012) Mycorrhiza in tropical crops. Contribution to agricultural food sustainability. *Editorial Académica Española*, pp. 1-239.

Ruiz M, Muñoz Y, Dell'Amico J, Cabrera A, Simó J (2016) Evaluation of different strains of arbuscular mycorrhizae in the development of rice plant (*Oryza sativa* L.) in flooded soil conditions. *Cultivos Tropicales,* 37(4): 67-75.

Ruiz M, Valdés AR, Herrera-Peraza R (1999) Differential functions of several strains of *Glomus* under pH changes. *Acta Botánica Cubana,* pp. 1-133.

Ruiz-Lozano J, Porcel R, Bárzana G, Azcón R, Aroca R (2012) Contribution of arbuscular mycorrhizal symbiosis to plant drought tolerance: state of the Art. In: Aroca R (Eds.) *Plant Responses to Drought Stress.* Springer, Berlin, Heidelberg, pp. 335-362. doi:10.1007/978-3-642-32653-0_13.

Ryan MH, Graham JM, Morton JB, Kirkegaard JA (2019) Research must use a systems agronomy approach if management of the arbuscular mycorrhizal symbiosis is to contribute to sustainable intensification. A response to Rillig et al. (2018) 'Why farmers should manage the arbuscular mycorrhizal symbiosis'. *New Phytologist,* 222(3):1176-1178. doi:10.1111/nph.15600.

Sánchez C, Caballero D, Rivera R, Cupull R (2006a) Response of mycorrhizal fungi (HMA) strains to coffee seedling development (Part I) Gleyic Cambisol soil. *Centro Agrícola,*33(1):33-38.

Sánchez C, Caballero D, Rivera R. Cupull R, González C, Ferrer M, Delgado Y(2006b) Response of mycorrhizal fungi (HMA) strains to coffee seedling development (Part III) . Rhodic Eutric Cambisol soil. *Centro Agrícola,* 33(2):11-16.

Sánchez C, Caballero D, Rivera R. Cupull R, González C, Ferrer M, Delgado Y(2006c) Response of mycorrhizal fungi (HMA) strains to coffee seedling development (Part II) Lixiviated Red Ferralitic soil. *Centro Agrícola,* 33(2):17-22.

Sánchez C, Rivera R, Caballero D, Cupull R, González C, Urquiaga-Caballero S (2011) Green manures and mycorrhizal inoculation of coffee seedlings on Lixiviated Red Ferralitic soil. *Cultivos Tropicales,* 32(3):11-17.

Sánchez C, Rivera R, González C, Cupull R, Herrera R, Varela M (2000) Effect of mycorrhizal fungi inoculation on the production of coffee plant seedlings in three soils of Guamuhaya Mountains. *Cultivos Tropicales,* 21(3):5-13.

Santos-González JC, Nallanchakravarthula S, Alström S, Finlay RD (2011) Soil, but not cultivar, shapes the structure of arbuscular mycorrhizal fungal assemblages associated with strawberry. *Microbial Ecology,* 62:25-35. doi: 10.1007/s00248-011-9834-7.

Schütz L, Gattinger A, Meier M, Müller A, Boller T, Mäder P, Mathimaran N (2018) Improving crop yield and nutrient use efficiency via biofertilization-A global meta-analysis. *Frontiers in Plant Science,* 8:2204. doi: 10.3389/fpls.2017.02204.

Shüßler A, Walker C (2011) *The Glomeromycota: a species list with new families and new genera.* Gloucester: The Royal Botanic Garden Edinburgh, The Royal Botanic Garden Kew, BotanischeStaatssammlung Munich, and Oregon State University, pp. 1-58.

Sieverding E (1991) Vesicular Arbuscular Mycorrhiza in Tropical Agrosystem. German Agency for Technical Cooperation. *GTZ, GMBH*, Federal Republic of Germany, pp. 1-371.

Sieverding E, da Silva G A, Berndt R, Oehl F (2014) *Rhizoglomus*, a new genus of the *Glomeraceae*. *Mycotaxon,* 129(2):373-386. doi:10.5248/129.373.

Sikes BA, Maherali H, Klironomos JN (2014) Mycorrhizal fungal growth responds to soil characteristics, but not host plant identity, during a primary lacustrine dune succession. *Mycorrhiza,* 24:219-226. doi: 10.1007/s00572-013-0531-x.

Simó J, Rivera R, Ruiz L (2018) Application of arbuscular mycorrhizal fungi in tomato cv. 'INIVIT-T-2007" *Solanum lycopersicum* L. *Rev Agricultura Tropical,* 4(1):7-16.

Simó J, Rivera R, Ruiz L, Martín G (2020) The integration of AMF inoculants, green manure and organo-mineral fertilization, in banana plantations on Eutric Cambisol soil. *Tropical and Subtropical Agroecosystems,* 23:2882.

Simó J, Rivera R, Ruiz L, Ruiz M, Diaz G (2019) Effectiveness of arbuscular mycorrhizal fungi inoculated on *Canavalia ensiformis* L. in Calcaric Phaozem. *Agronomia Mesoamericana,* 30 (2):395-405. doi:10.15517/am.v30i2.33221

Simó JE, Ruiz L, Rivera R (2015) Management of arbuscular mycorrhizal symbiosis and nutrient supply in banana plantations planted with "FHIA-18" on Eutric Cambisol soil. *Cultivos Tropicales,* 36(4):43-54.

Simó JE, Ruiz LA, Rivera R (2017) Inoculation of arbuscular mycorrhizal fungi (AMF) and soil Cambisol - organic fertilizers ratio in the acclimatization of banana vitroplants. *Cultivos Tropicales,* 38 (3):102-111.

Siqueira JO, Collozzi-Filhho A, Saggin-Junior OJ, Guimaraes PTG, Oliveira E (1993) Seedling growth and coffee production under the influence of mycorrhizal fungi and superphosphate. *Rev Bras Cienc Solo,* 17(1):53-60.

Siqueira JO, Franco AA (1988) *Soil Biotechnology. Basis and prospect.* MEC-ESAL-FAEPE-ABEAS, Brasilia, D.F., pp. 1-235.

Smith SE, Smith FA (2011) Roles of arbuscular mycorrhizas in plant nutrition and growth: New paradigms from cellular to ecosystem scales. *Annual Review of Plant Biology,* 62:227-250. doi: 10.1146/annurev-arplant-042110-103846.

Stewart WM (2007) Considerations for the effective use of nutrients. *Informaciones Agronómicas,*67:1–7.

Tamayo Y, Martín GM, Corona Y, Barraza FV (2015) Response of the *Canavalia ensiformis* (L) D.C. co-inoculation with Rhizobium and arbuscular mycorrhizal fungi. *Hombre, Ciencia y Tecnología,* 19(1):100-108.

Tamayo-Aguilar Y, Martín-Alonso GM, Herrera-Altuve JA, Cesar-Gainza A, Abad-Michael M, Nápoles-García MC, Rivera R, Juárez-López P (2021) Biofertilizers in the succession *Canavalia ensiformis – Solanum lycopersicum*: yield and quality in tomato fruit. Revista *Fitotecnia Mexicana,* 44 (3) :341-347. doi: 10.35196/rfm. 2021.3.341.

The Royal Society (2009) *Reaping the benefits: science and the sustainable intensification of global agriculture*. The Royal Society Policy Document 11/09. The Royal Society, London, pp. 1-150.

Thirkell TJ, Cameron DD, Hodge A (2016) Resolving the 'nitrogen paradox' of arbuscular mycorrhizas: fertilization with organic matter brings considerable benefits for plant nutrition and growth. *Plant, Cell & Environment,* 39(8):1683-1690. doi: 10.1111/pce. 12667.

Tilman D, Cassman KG, Matson PA, Naylor R, Polasky S (2002) Agricultural sustainability and intensive production practices. *Nature,* 418:671-677. doi:10.1038/ nature01014.

Tobacco Research Institute (2012) *Technical Guidance on Tobacco Cultivation in Cuba.* Editor. Ministry of Agriculture – Tobacco Research Institute, Artemisa, Cuba, pp. 1-148.

van der Heijden MGA, Martin FC, Selosse MA, Sanders IR (2015) Mycorrhizal ecology and evolution: the past, the present, and the future. *New Phytologist,* 205:1406-1423. doi:10.1111/nph.13288.

Vanlauwe B, Bationo A, Chianu J, Giller KE, Merckx R, Mokwunye U, Ohiokpehai O, Pypers P, Tabo R, Shepherd KD, Smaling EMA, Woomer PL, Sanginga N (2010) Integrated soil fertility management: Operational definition and consequences for implementation and dissemination. *Outlook on Agriculture*, 39: 17–24. doi:10.5367/ 000000010791169998.

Vargas T. de O, Diniz ER, Pacheco ALV, Silva RH, Urquiaga S (2017) Green manure-^{15}N absorbed by broccoli and zucchini in sequential cropping. *Scientia Horitculturae*, 214: 209-213. doi:10.101 6/j.scienta.2016.11.028.

Volf MR, Guimarães TM, Scudeletti D, Cruz IV, Rosolem CA (2018) Potassium dynamics in Ruzigrass rhizosphere. *Rev Bras Cienc Solo,* 42:e0170370.doi:10.1590/18069657 rbcs20170370.

Willis A, Rodrigues BF, Harris PJC (2013) The ecology of arbuscular mycorrhizal fungi. *Critical Reviews in Plant Sciences,* 32:1–20. doi: 10.1080/07352689.2012.683375.

WRB (World reference base for soil resources) (2014) International soil classification system for naming soils and creating legends for soil maps. *World Soil Reports* no. 106. FAO, Roma, 81 p.

Yang C, Ellouze W, Navarro-Borrell A, Esmaeili Taheri A, Klabi R, Dai M, Kabir Z, Hamel C (2014) Management of the arbuscular mycorrhizal symbiosis in sustainable crop production.In: Solaiman Z, Abbott L, Varma A (Eds.) Mycorrhizal Fungi: Use in Sustainable Agriculture and Land Restoration. *Soil Biology*, vol 41. Springer, Berlin, Heidelberg, pp. 89-118.doi: 10.1007/978-3-662-45370-4_7.

Yasmeen T, Hameed S, Tariq M, Ali S (2012) Significance of arbuscular mycorrhizal and bacterial symbionts in a **tripartite** association with *Vigna radiate. Acta Physiologiae Plantarum,*34:1519-1528. doi.org/10.1007/s11738-012-0950-x

Zhang S, Lehmann A, Zheng W, You Z, Rillig MC (2018) Arbuscular mycorrhizal fungi increase grain yields: a meta-analysis.*New Phytologist,* 222(1):543-555. doi: 10.11 11/nph.15570.

Chapter 5

Arbuscular Mycorrhizal Fungi Assisted Phytoremediation and Its Role in Sustainable Soil Management

Ranjith Sellappan[1,*]
Anandkumar Selvaraj[1]
Senthamilselvi Dhandapani[1]
Aswini Krishnan[2]
Kasthuri Selvakumar[1]
and Kalaiselvi Thangavel[1]
[1] Department of Agricultural Microbiology,
Tamil Nadu Agricultural Universtiy, Coimbatore, Tamil Nadu, India
[2] Division of Microbiology,
Indian Agricultural Research Institute, Pusa, New Delhi, India

Abstract

Rapid urbanisation and industrialization have resulted in a serious problem of polluted soil contamination, which causes distress to living organisms and ecosystem imbalance. Hence, there is an immediate global demand for the diminution of contaminants in polluted soil without causing a negative effect on the environment. Bioremediation is the most efficient and promising alternative strategy employed to decontaminate soil pollutants using different organisms, including microorganisms.

Over time, microbial remediation processes have been accelerated to produce better, more eco-friendly, and biodegradable materials for the

* Corresponding Author's Email: s.ranjith994@gmail.com.

In: New Research on Mycorrhizal Fungus
Editors: Qiang-Sheng Wu, Ying-Ning Zou, Yue-Jun He et al.
ISBN: 979-8-88697-637-3

complete dissemination of the contaminants. Arbuscular mycorrhizal (AM) fungi are ubiquitous soil microorganisms that form symbiotic associations with most of the land plants, which improve plant nutrition, water relations, plant productivity and diversity, ecosystem establishment, and plant protection against biotic and abiotic environmental stress. Despite improving plant growth and health, AM fungi also improve soil health, which includes mycorrhizal remediation of soil contaminants. AM fungi-assisted bioremediation not only removes the soil pollutants but also enhances the host plant's growth and development by regulating tolerance mechanisms.

This chapter discusses the role of AM fungi in soil bioremediation and pollutant tolerance in plants, as well as advancements in AM fungi-assisted remediation that have contributed to sustainable soil management.

Keywords: phytoremediation, mycorrhiza, soil, bioremediation, pollutant, soil health

1. Introduction

Soil contamination is defined as undesirable changes in the physical, chemical, and biological nature of soil that cause harm to humans, animals, plants, and ecosystems (Brevik et al., 2020). Anthropogenic activity such as the usage of pesticides, herbicides, industrial dyes, and chemicals was affecting soil quality and health (Lopez-Pacheco et al., 2019). The effect of anthropogenic activity on soil contamination is becoming a severe problem.

The main sources of soil pollution are inorganic fertiliser usage in agricultural fields, industrial effluents, and urbanisation activities. Important components of soil pollution are inorganic chemicals, heavy metals, pesticides, and solvents. Soil contamination causes reduced soil fertility and health, which are directly linked with crop production and an imbalance in soil flora and fauna (Gunstone et al., 2021). To alleviate this global problem, we should find eco-friendly remediation methods.

Plants have undergone various environmental stresses (biotic and abiotic) during their growth period. Plants have their own mechanisms to mitigate and/or tolerate these stresses (Waqas et al., 2019). Phytoremediation is defined as plant-based mitigation of soil pollutants through the following mechanisms (Figure 1).

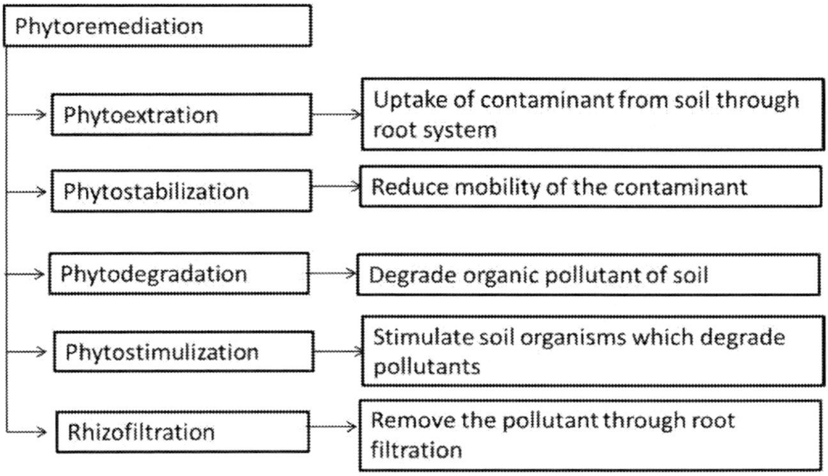

Figure 1. Phytoremediation mechanisms.

Besides, plant-associated microbes also play an important role in the remediation of soil pollutants. Plant root-associated rhizosphere microbes play a pivotal role in the remediation of soil pollutants. Arbuscular mycorrhizal (AM) fungi (AMF) are ubiquitous soil microbes that have mutualistic associations with more than 80% of plant species. Mycorrhizae also play an important role in the remediation of soil contaminants. The phytoremediation efficacy in soil was also improved by AMF's symbiotic relationship with plants (Yang et al., 2016). In this chapter, we address the importance of arbuscular mycorrhizal fungi in phytoremediation for long-term agricultural output.

2. AMF and Plant Mutualism

AM fungi are the most prevalent mycorrhizal fungi, and their distribution is the most widespread. Most terrestrial plants have AM fungi as root symbionts, which are ecologically significant because they connect biotic and abiotic ecosystem components in the soil via carbon and nutrient fluxes (Chourasiya et al., 2021).The degree of colonisation of roots by AM fungus varies depending on the kind of interplay between host, symbiont, and environment. Despite the fact that many researchers have undertaken studies on the natural associations of AM fungus to discover their variety in various types of plants during the last few years, The presence or absence of AM fungal spores indicates that these species may be active in the soil at a given time, colonising and proliferating in the plant's host roots. Because AMF is not host-specific, many AMF species may colonise a single host plant, and a single fungus may be linked to multiple host plants. Once AMF has colonised the roots, plants appear to be able to control additional colonization, and seedlings become mycorrhizal quickly if the soil is rich in mycorrhizal hyphae (Wipf et al., 2019).

AMF's association with plants was first documented 4 billion years ago. Mycorrhizal fungi enhance the nutritional status of their host plants by regulating mineral nutrition, water absorption, development, and disease resistance, while the host plant is required for fungal growth and reproduction (Bhantana et al., 2021). Mycorrhizal symbioses may be divided into various categories at the most basic level, including arbuscular mycorrhizal and ectomycorrhizal, which have different evolutionary origins (Van der Heijden et al., 2015). The presence of a host is detected by AM fungus via root-exuded molecules, which cause spore germination and hyphal branching (Lucini et al., 2019). Carotenoid-derived phytohormones termed strigolactones (SLs), which play a key role in plant growth, are the most researched plant symbiotic signals (Borghi et al., 2016). AM fungi also produce signal molecules that cause plant symbiotic responses such as transcriptional activation, nucleus-associated Ca^{2+} signals, root starch storage, and lateral root formation. Plant signalling pathways involved in the perception of AM fungal signals have been studied in legumes like *M. truncatula*, partly as a result of studies into rhizobial Nod factor signalling (Koziol et al., 2016). Mycorrhizal fungi are classified into two groups: aseptate endophytes like Glomeromycota and septate Asco- and Basidiomycota. Mycorrhizal interactions are usually divided into two major types depending on plant anatomical characteristics: hyphae stay extracellular in ectomycorrhiza, which colonise plants and shrubs and generate significant alterations in root development, while their presence causes minor changes in epidermal or cortical cells. Hyphae penetrate root cells to develop an intracellular symbiosis in endomycorrhiza, such as ericoid, orchid, and arbuscular mycorrhizae (Bhat et al., 2016).

3. AMF Impart Plant Tolerance against Soil Contaminants

Soil, in its native form, is an uncontaminated material covering the earth's surface, supporting the life of other organisms. As time passes, anthropogenic activities have introduced several chemical substances, which later became a threat to the natural soil ecosystem in the form of contaminants (Adriano, 2001). Most common soil contaminants include pesticides, excessive fertilizers, heavy metals, polycyclic or polynuclear aromatic hydrocarbons, etc. These pollutants are posing severe threats to the soils and their associated life forms, especially plants (Rodrguez-Eugenio et al., 2018). Insignificant soil physico-chemical properties such as deficient nutrient availability, extreme pH, poor structure, and hydraulic properties are the characteristics of contaminated soils, which in turn obstruct the plant's growth and development. The soil contaminants control plant growth and productivity by influencing plant physiology and morphology, interfering with the availability of essential nutrients, and worsening the activity of symbiotic plant growth-promoting microbes (Mittler, 2006; Miransari, 2016).

Several microbial symbionts are coupled with the plants and function together as "holobionts" to enhance their growth and survival (Vandenkoornhuyse et al., 2015; Sánchez-Caizares et al., 2017). Furthermore, mycorrhizal fungi go above and beyond to help plants by increasing their tolerance to various stress conditions (Cabral et al., 2015; Firmin et al., 2015; Neuenkamp et al., 2019), and AMF inoculation has been proposed as a viable biotechnology tool (Schneider et al., 2017) with significant advantages over other microorganisms. Above all, the widely spread mycelial structure of mycorrhizas translocates the trace elements and sequesters them as inaccessible complex chemical substances, thereby restraining their toxicity to plants (Colpaert et al., 2011; Taylor and Sinsabaugh, 2015; Shi et al., 2019). Instead, there are various other interaction mechanisms by which mycorrhiza enhances plant growth by improving essential nutrient availability, reducing the availability of heavy metals, transforming metals into less toxic forms, and alleviating metal-induced oxidative stress (Mishra et al., 2017). For instance, *Rhizophagus intraradices* modified the rhizospheric bacterial community by limiting Cd availability to rice (Chen et al., 2019).

Soil, in its native form, is an uncontaminated material covering the earth's surface, supporting the life of other organisms. As time passes, anthropogenic activities have introduced several chemical substances, which later became a threat to the natural soil ecosystem in the form of contaminants (Adriano, 2001). Most common soil contaminants include pesticides, excessive fertilizers, heavy metals, polycyclic or polynuclear aromatic hydrocarbons, etc. These pollutants are posing severe threats to the soils and their associated life forms, especially plants (Rodrguez-Eugenio et al., 2018). Insignificant soil physico-chemical properties such as deficient nutrient availability, extreme pH, poor structure, and hydraulic properties are the characteristics of contaminated soils, which in turn obstruct the plant's growth and development. The soil contaminants control plant growth and productivity by influencing plant physiology and morphology, interfering with the availability of essential nutrients, and worsening the activity of symbiotic plant growth-promoting microbes (Mittler, 2006; Miransari, 2016).

Several microbial symbionts are coupled with the plants and function together as "holobionts" to enhance their growth and survival (Vandenkoornhuyse et al., 2015; Sánchez-Caizares et al., 2017). Furthermore, mycorrhizal fungi go above and beyond to help plants by increasing their tolerance to various stress conditions (Cabral et al., 2015; Firmin et al., 2015; Neuenkamp et al., 2019), and AMF inoculation has been proposed as a viable biotechnology tool (Schneider et al., 2017) with significant advantages over other microorganisms. Above all, the widely spread mycelial structure of mycorrhizas translocates the trace elements and sequesters them as inaccessible complex chemical substances, thereby restraining their toxicity to plants (Colpaert et al., 2011; Taylor and Sinsabaugh, 2015; Shi et al., 2019). Instead, there are various other interaction mechanisms by which mycorrhiza enhances plant growth by improving essential nutrient availability, reducing the availability of heavy metals, transforming metals into less toxic forms, and alleviating metal-induced oxidative stress (Mishra et al., 2017). For instance, *Rhizophagus intraradices* modified the rhizospheric bacterial community by limiting Cd availability to rice (Chen et al., 2019).

The symbiotic relationship between plant roots and AMF allows photosynthetic products to pass from plants to fungi.Consecutively, the fungal symbiont supports plants by promoting plant nutrient uptake, enhancing photosynthesis, and accumulating biomass, in addition to reducing the toxicity of heavy metals (Chen et al., 2017; Mitra et al., 2020). The fungal hyphae penetrate into the root cortical cells and form specialised structures called arbuscules and vesicles (Verma et al., 2019), which prevent the translocation of heavy metals to the aerial plant parts by immobilising them inside the cortical cells. This, in turn, protects the leaf tissues from the accumulation of heavy metals, which restrict stomatal conductance and photosynthesis (Schubler, 2002). Application of mycorrhiza, including *Rhizophagus intraradices*, *Glomus versiforme,* and *Funneliformis mosseae* (Wu et al., 2016), substantially reduced the Pb concentration in Pakchoi shoots. In line with this, AMF increases the uptake of essential elements like Mg and Fe by promoting the de novo synthesis of certain proteins, wherein they constitute an indispensable role in the chlorophyll molecule (Abd_Allah et al., 2015) and enhances the photosynthetic ability and phytochemical efficiency of the plants.

On the contrary, AMF also renders the soil contaminants incapable by precipitating and chelating them in the rhizosphere, sequestering them in vacuoles, activating the antioxidant mechanisms of plants, and immobilising them in mycelial or hyphal structures (Mishra et al., 2019). The vesicles of mycorrhizal fungi are analogous to the plant vacuoles, where toxic compounds would accumulate. Just after the metal contaminants came into contact with the target site, the polyphosphate granules triggered the sequestration process (Abdelhameed and Metwally, 2019; Shi et al., 2019), which in turn acquiesced in a higher concentration of metal contaminants in the aboveground plant parts without any phytotoxic effects. The hyphal structure and mycelium present inside and outside the plant roots act as biological filters, thereby interfering with the uptake of toxic metals and retaining those elements in the roots. Furthermore, the number of vesicles increases with an increase in the concentration of toxic compounds (Yang et al., 2015). Subsequently, under certain circumstances, the biosynthesis of plant cell walls is enhanced by the AMF (Zhang et al., 2019), so that the increase in thickness prevents the entry of heavy metals through the cell wall, ultimately minimising the accumulation of metal ions inside the plant cells.

Likewise, under stressed conditions, AMF promotes plant growth by altering the soil structure. The fungal hyphae take up nutrients from the soil and translocate them to the fungus-plant interface, where they get exchanged bidirectionally (Janeeshma and Puthur, 2020). Chiefly, the phosphate transporters of mycorrhizal fungi regulate the movement of inorganic phosphate from soil to plants (Johri et al., 2015). Also, the transporter proteins such as GmosPT, GiPT, and GvPT are present on the extraradial mycelium of *Glomus mossae*, *G. intraradices,* and *G. versiforme*, which sustain the P uptake by plants (Benedetto et al., 2005; Maldonado-Mendoza et al., 2001; Harrison and Buuren, 1995).

Besides other mechanisms, AMF produces glomalin-related soil proteins (GRSPs), which are glycoproteins analogous to Hsp60 (heat shock protein 60) (Wright and Upadhyaya, 1998). Glomalin produced by the AMF helps the plants to tolerate soil contaminants by improving the soil structure and reducing the availability of metal compounds by forming protein–metal complexes in the soil that restrict their entry into the plants (Chern et al., 2007; Gonzalez-Chavez et al., 2009). Moreover, several soil enzymes are modulated by the activity of different AMF species (Wu et al. 2015), which play an imperative role in nutrient cycling, stabilising soil structure, etc. (Burke et al.

2011). Hence, it has been widely believed that AMF tolerance provides tolerance to host plants against several stress conditions such as salinity, drought, soil contaminants, and so on.

4. AMF Mediated Phytoremediation in Pesticide and Heavy Metal Contaminated Soil

Pesticides are synthetic organic chemicals used to control undesirable or hazardous plants, insects, and pests that feed on crops. Pesticides are classified according to the target they are meant to kill (e.g., insecticides and herbicides control insects and weeds, respectively). Herbicides rank first among commonly used pesticides in the United States. They range from non-selective to highly selective, with post-emergence, pre-plant, and pre-emergence applications, and are used to control certain weeds in specific crops. Insecticides and fungicides are the next most commonly used chemicals. Intensive usage of these chemicals over the last few decades has spoiled the environment and polluted the soil. The pesticide and heavy metal residues in soil can adversely impact beneficial plants and organisms, as is well documented (He et al., 2005). Pesticide residues persist in soils, where they might be absorbed by soil particles or leached into ground water. Because of their mobility and solubilities, human exposure to these contaminants through the intake of contaminated food or drinking water can lead to their accumulation in humans, plants, and animals and also cause significant damage to the environment.

The nature of chemical residues in soil is influenced by a number of soil components, which influences their availability to plants (Boruvka and Drabek, 2004). According to Kapoor and Viraraghavan (1995), the free amino, hydroxyl, carboxyl, and other groups in the cell wall of soil fungi can bind to potentially hazardous substances like Cu, Pb, Cd, etc. Many filamentous fungi can absorb and use these trace elements in their commercial biosorbants (Morley and Gadd, 1995). Likewise, the proteins present in the cell walls of AM fungi have similar abilities to absorb potentially hazardous substances in the soil by sequestration. One of these components is glomalin, a glycoprotrien produced by the hyphae of AMF fungi (Wright and Upadhyaya, 1998), which is released into soil by AMF hyphae. Gonzalez-Chavez et al. (2004) showed that glomalin secreted on the hyphae of AMF plays a vital role in the sorption and sequestration of potentially hazardous substances, which lowers their bioavailability in soil. AMF with different strategies showed higher tolerance to these organic pollutants and heavy metals, according to numerous reports (Salam et al., 2017). These strategies include morphological adaptation, protective molecules (antioxidants, chaperones, and trehalose), pollutant transport, compartmentation, transformation, and changes in gene expression. However, knowledge about the mechanisms of dealing with pesticides is limited because the studies on AMF and pesticides are mostly descriptive rather than mechanistic (Ahmed et al., 2018). Mycorrhizal associations could greatly enhance host plant HM tolerance by improving plant nutrient acquisition and by influencing the fate of the metals in both the plants and soil (Miransari, 2011). *G. intraradices* Sy167 strain was found to be effective in alleviating heavy metal stress in plants (Hildebrandt et al., 1999; Kaldorf et al., 1999). *Thlaspi* sp. colonised by *G. intraradices is* reported to be well adapted to heavy metal-polluted soil types (Regvar et al., 2003).

AMF play an important ecological role in the phytostabilization of potentially harmful trace elements in polluted soils via sequestration, allowing mycorrhizal plants to survive in polluted soils. Several studies have focused on the diversity and tolerance of AMF in heavy metal-contaminated soils to understand the adaptation and tolerance of AMF to heavy metals in soils, since this could facilitate the management of AM for restorational bioremediation programs.

5. Role of AMF in Soil Health Improvement in Sustainable Agriculture

Phytoaccumulation/phytoextraction, or the employment of plants to clean up non-volatile hydrocarbons and immobile inorganic pollutants in polluted soils, has shown promise as a new strategy for the in situ clean-up of huge volumes of low- to moderately contaminated soils in recent years. Mycorrhizal application may successfully reduce the quantitative usage of chemical fertiliser input, especially phosphorus, and it is commonly assumed that AMF might be regarded as an alternative to inorganic fertilisers in the near future (Ortas, 2012).

"Fungal genotype, absorption of metal by plant via AM symbiosis, root length density, competition between AMF communities, the rhizosphere (pH, CEC, etc.), the metal itself, concentrations of accessible metals, soil contamination conditions, interactions between P and metals, experimental condition" (Giasson et al., 2008). Recent investigations on live cell imaging of the peri-arbuscular membrane have revealed two domains: the arbuscule branch domain, which contains the phosphate transporter MtPT4 unique to symbiosis, and the arbuscule trunk domain, which contains the copper binding protein MtBcp1 (Pumplin and Harrison, 2009). Arginine has been implicated in the transfer of nitrogen to both fungal and plant cells during periods of ammonium feeding and is eventually absorbed by the fungal systems through NH_4^+ assimilation. The soil microbiome has been found to be beneficial in limiting the negative impacts of abiotic stresses in recent studies, although the mechanism is still unknown. Furthermore, osmotic changes mediated by the accumulation of substances such as proline, chlorophyll, and carotenoids are another impact of AMF-induced drought tolerance (Abbaspour et al., 2012).

The thick network of AMF mycelia provides a larger surface area, which improves nutrient intake by plant roots and facilitates soil aggregation (Lenoir et al., 2016). AMF can improve the nutritional content of crops by influencing and increasing the synthesis of carotenoids and other volatile chemicals (Hart et al., 2015). Attempts are being made to investigate potential ways to improve crop productivity in salt-affected soils. The careful application of AMF for alleviating salinity-induced detrimental effects on plants is one such potential strategy (Santander et al., 2019). Under aluminium stress, the interaction of AMF with wheat boosted nutrient absorption (Aguilera et al., 2014). The efficacy and mechanism of phytoremediation using the woody legume *Robinia pseudoacacia* L. infected with AMF are examined in the current study. Because of its quick growth, high biomass, ability to accumulate huge quantities of HM, and atmospheric nitrogen (N) fixation, *R. pseudoacacia* has been claimed to have the potential to remove metal pollutants from soil (Yang et al., 2015). Heavy metal chelation might be aided by the chemicals generated by the extraradical mycelium. AMF has been proven to be damaged by heavy metal toxicity; however, mycotrophic plants growing in heavy metal-contaminated soils are frequently colonised by AMF (Leyval et al., 1997). However, additional field research is needed to uncover the critical functions of AMF in phytoremediation and ecological stability on larger scales and in more complicated environments.

Conclusion

Anthropogenic activity has resulted in global soil contamination. Besides opting for various chemical and physical methods to clean up those contaminants, phytoremediation is a potential option for ecofriendly management of soil health. Additionally, some endophytic microbes, such as mycorrhiza, play an important role in soil decontamination. Mycorrhiza has the capability of uptaking and/or degrading complex inorganic chemical pollutants in soil. Mycorrhiza could be the best source for mitigating soil contamination and health, which pave the way for sustainable agricultural production.

References

Abbaspour H, Saeidi-Sar S, Afshari H and Abdel-Wahhab MA (2012) Tolerance of mycorrhiza infected pistachio (*Pistacia vera* L.) seedling to drought stress under glasshouse conditions. *Journal of Plant Physiology*, 169(7): 704-709.

Abd Allah EF, Abeer H, Alqarawi AA and Hend AA (2015) Alleviation of adverse impact of cadmium stress in sunflower (*Helianthus annuus* L.) by arbuscular mycorrhizal fungi. *Pakistan Journal of Botany*, 47(2): 785-795.

Abdelhameed RE and Metwally RA (2019) Alleviation of cadmium stress by arbuscular mycorrhizal symbiosis. *International Journal of Phytoremediation*, 21(7): 663-671.

Adriano DC (2001). *Trace Elements in Terrestrial Environments: Biogeochemistry, Bioavailability, and Risks of Metals* (Vol. 860). New York: Springer.

Aguilera P, Cornejo P, Borie F, Barea JM, Von Baer E and Oehl F (2014) Diversity of arbuscular mycorrhizal fungi associated with *Triticum aestivum* L. plants growing in an Andosol with high aluminum level. *Agriculture, Ecosystems & Environment*, 186: 178-184.

Benedetto A, Magurno F, Bonfante P, and Lanfranco L (2005) Expression profiles of a phosphate transporter gene (GmosPT) from the endomycorrhizal fungus *Glomus mosseae*. *Mycorrhiza*, 15(8): 620-627.

Bhantana P, Rana MS, Sun XC, Moussa MG, Saleem MH, Syaifudin M and Hu CX (2021) Arbuscular mycorrhizal fungi and its major role in plant growth, zinc nutrition, phosphorous regulation and phytoremediation. *Symbiosis*, 84(1): 19-37.

Bhat RA, Haq S, Dervash MA, Bhatti AA, Nissa M and Mir MR (2016) Arbuscular mycorrhizal fungi boon for plant nutrition and soil health. In *Soil Science: Agricultural and Environmental Prospectives* (pp. 317-332). Springer, Cham.

Borghi L, Liu GW, Emonet A, Kretzschmar T and Martinoia E (2016) The importance of strigolactone transport regulation for symbiotic signaling and shoot branching. *Planta*, 243(6): 1351-1360.

Boruvka L and Drábek O (2004) Heavy metal distribution between fractions of humic substances in heavily polluted soils. *Plant Soil and Environment*, 50(8): 339-345.

Brevik EC, Slaughter L, Singh BR, Steffan JJ, Collier D, Barnhart P and Pereira P (2020) Soil and human health: current status and future needs. *Air, Soil and Water Research*, 13.

Burke DJ, Weintraub MN, Hewins CR and Kalisz S (2011) Relationship between soil enzyme activities, nutrient cycling and soil fungal communities in a northern hardwood forest. *Soil Biology and Biochemistry*, 43(4): 795-803.

Cabral L, Soares CRFS, Giachini AJ and Siqueira JO (2015) Arbuscular mycorrhizal fungi in phytoremediation of contaminated areas by trace elements: mechanisms and major benefits of their applications. *World Journal of Microbiology and Biotechnology*, 31(11): 1655-1664.

Chen S, Zhao H, Zou C, Li Y, Chen Y, Wang Z and Ahammed GJ (2017) Combined inoculation with multiple arbuscular mycorrhizal fungi improves growth, nutrient uptake and photosynthesis in cucumber seedlings. *Frontiers in Microbiology*, 8: 2516.

Chen XW, Wu L, Luo N, Mo CH, Wong MH and Li H (2019) Arbuscular mycorrhizal fungi and the associated bacterial community influence the uptake of cadmium in rice. *Geoderma*, 337: 749-757.

Chern EC, Tsai DW and Ogunseitan OA (2007) Deposition of glomalin-related soil protein and sequestered toxic metals into watersheds. *Environmental Science & Technology*, 41(10): 3566-3572.

Chourasiya D, Gupta MM, Sahni S, Oehl F, Agnihotri R, Buade R and Sharma MP (2021) Unraveling the AM fungal community for understanding its ecosystem resilience to changed climate in agroecosystems. *Symbiosis*, 84(3): 295-310.

Colpaert JV, Wevers JH, Krznaric E and Adriaensen K (2011) How metal-tolerant ecotypes of ectomycorrhizal fungi protect plants from heavy metal pollution. *Annals of Forest Science,* 68(1): 17-24.

Firmin S, Labidi S, Fontaine J, Laruelle F, Tisserant B, Nsanganwimana F and Sahraoui ALH. (2015) Arbuscular mycorrhizal fungal inoculation protects Miscanthus× giganteus against trace element toxicity in a highly metal-contaminated site. *Science of the Total Environment,* 527: 91-99.

Giasson P, Karam A and Jaouich A (2008) Arbuscular mycorrhizae and alleviation of soil stresses on plant growth. *Mycorrhizae: Sustainable Agriculture and Forestry*, 99-134.

Gonzalez-Chavez MC, Carrillo-Gonzalez R and Gutierrez-Castorena MC (2009) Natural attenuation in a slag heap contaminated with cadmium: the role of plants and arbuscular mycorrhizal fungi. *Journal of Hazardous Materials*, 161(2-3): 1288-1298.

Gunstone T, Cornelisse T, Klein K, Dubey A and Donley N (2021) Pesticides and soil invertebrates: A hazard assessment. *Frontiers in Environmental Science*, 9: 643847.

Hage-Ahmed K, Rosner K and Steinkellner S (2019) Arbuscular mycorrhizal fungi and their response to pesticides. *Pest Management Science*, 75(3): 583-590.

Harrison MJ and Buuren MLV (1995) A phosphate transporter from the mycorrhizal fungus Glomus versiforme. *Nature*, 378(6557): 626-629.

Hart M, Ehret DL, Krumbein A, Leung C, Murch S, Turi C and Franken P (2015) Inoculation with arbuscular mycorrhizal fungi improves the nutritional value of tomatoes. *Mycorrhiza*, 25(5): 359-376.

He ZL, Yang XE and Stoffella PJ (2005) Trace elements in agroecosystems and impacts on the environment. *Journal of Trace Elements in Medicine and Biology*, 19(2-3): 125-140.

Hildebrandt U, Kaldorf M and Bothe H (1999) The zinc violet and its colonization by arbuscular mycorrhizal fungi. *Journal of Plant Physiology*, 154: 709-717.

Hildebrandt U, Kaldorf M and Bothe H (1999) The zinc violet and its colonization by arbuscular mycorrhizal fungi. *Journal of Plant Physiology*, 154(5-6): 709-717.

Janeeshma E and Puthur JT (2020) Direct and indirect influence of arbuscular mycorrhizae on enhancing metal tolerance of plants. *Archives of Microbiology*, 202(1), 1-16.

Johri AK, Oelmueller R, Dua M, Yadav V, Kumar M, Tuteja N and Stroud RM (2015) Fungal association and utilization of phosphate by plants: success, limitations, and future prospects. *Frontiers in Microbiology*, 6: 984.

Kaldorf M, Kuhn AJ, Schröder WH, Hildebrandt U and Bothe H (1999) Selective element deposits in maize colonized by a heavy metal tolerance conferring arbuscular mycorrhizal fungus. *Journal of Plant Physiology*, 154(5-6): 718-728.

Kapoor A and Viraraghavan T (1995) Fungal biosorption - an alternative treatment option for heavy metal bearing wastewaters: a review. *Bioresource Technology*, 53(3): 195-206.

Koziol L and Bever JD (2016) AMF, phylogeny, and succession: specificity of response to mycorrhizal fungi increases for late-successional plants. *Ecosphere*, 7(11): e01555.

Lenoir I, Lounes-Hadj Sahraoui A and Fontaine J (2016) Arbuscular mycorrhizal fungal-assisted phytoremediation of soil contaminated with persistent organic pollutants: a review. *European Journal of Soil Science*, 67(5): 624-640.

Leyval C, Turnau K and Haselwandter K (1997) Effect of heavy metal pollution on mycorrhizal colonization and function: physiological, ecological and applied aspects. *Mycorrhiza*, 7(3): 139-153.

López-Pacheco IY, Silva-Núñez A, Salinas-Salazar C, Arévalo-Gallegos A, Lizarazo-Holguin, LA, Barceló D and Parra-Saldívar R (2019) Anthropogenic contaminants of high concern: existence in water resources and their adverse effects. *Science of the Total Environment,* 690: 1068-1088.

Lucini L, Colla G, Moreno MBM, Bernardo L, Cardarelli M, Terzi V adn Rouphael Y (2019) Inoculation of *Rhizoglomus irregulare* or *Trichoderma atroviride* differentially modulates metabolite profiling of wheat root exudates. *Phytochemistry*, 157: 158-167.

Maldonado-Mendoza IE, Dewbre GR and Harrison MJ (2001) A phosphate transporter gene from the extra-radical mycelium of an arbuscular mycorrhizal fungus *Glomus intraradices* is regulated in response to phosphate in the environment. *Molecular Plant-Microbe Interactions*, 14(10): 1140-1148.

Miransari M (2011) Hyperaccumulators, arbuscular mycorrhizal fungi and stress of heavy metals. *Biotechnology Advances*, 29(6): 645-653.

Miransari M (2016) Soybean production and heavy metal stress. In *Abiotic and Biotic Stresses in Soybean Production* (pp. 197-216). Academic Press.

Mishra A, Bhattacharya A and Mishra N (2019) Mycorrhizal symbiosis: an effective tool for metal bioremediation. In *New and Future Developments in Microbial Biotechnology and Bioengineering* (pp. 113-128). Elsevier.

Mishra J, Singh R and Arora NK (2017) Alleviation of heavy metal stress in plants and remediation of soil by rhizosphere microorganisms. *Frontiers in Microbiology*, 8: 1706.

Mitra D, Uniyal N, Panneerselvam P, Senapati A and Ganeshamurthy AN (2020) Role of mycorrhiza and its associated bacteria on plant growth promotion and nutrient management in sustainable agriculture. *International Journal of Life Sciences and Applied Sciences*, 1(1): 1-10.

Mittler R (2006) Abiotic stress, the field environment and stress combination. *Trends in Plant Science*, 11(1): 15-19.

Morley GF and Gadd GM (1995) Sorption of toxic metals by fungi and clay minerals. *Mycological Research*, 99(12): 1429-1438.

Neuenkamp L, Prober SM, Price JN, Zobel M and Standish RJ (2019) Benefits of mycorrhizal inoculation to ecological restoration depend on plant functional type, restoration context and time. *Fungal Ecology*, 40: 140-149.

Ortas I (2012) The effect of mycorrhizal fungal inoculation on plant yield, nutrient uptake and inoculation effectiveness under long-term field conditions. *Field Crops Research*, 125: 35-48.

Pumplin N and Harrison MJ (2009) Live-cell imaging reveals periarbuscular membrane domains and organelle location in *Medicago truncatula* roots during arbuscular mycorrhizal symbiosis. *Plant Physiology*, 151(2): 809-819.

Regvar M, Vogel K, Irgel N, Wraber T, Hildebrandt U, Wilde P and Bothe H (2003) Colonization of pennycresses (*Thlaspi* spp.) of the Brassicaceae by arbuscular mycorrhizal fungi. *Journal of Plant Physiology*, 160(6): 615-626.

Rodríguez-Eugenio N, McLaughlin M and Pennock D (2018) Soil pollution: a hidden reality. FAO.

Sanchez-Canizares C, Jorrín B, Poole PS and Tkacz A (2017) Understanding the holobiont: the interdependence of plants and their microbiome. *Current Opinion in Microbiology*, 38: 188-196.

Santander C, Sanhueza M, Olave J, Borie F, Valentine A and Cornejo P (2019) Arbuscular mycorrhizal colonization promotes the tolerance to salt stress in lettuce plants through an efficient modification of ionic balance. *Journal of Soil Science and Plant Nutrition*, 19(2): 321-331.

Schneider J, Bundschuh J, De Melo Rangel W and Guilherme LRG (2017) Potential of different AM fungi (native from As-contaminated and uncontaminated soils) for supporting Leucaena leucocephala growth in As-contaminated soil. *Environmental Pollution*, 224: 125-135.

Schubler A (2002) Molecular phylogeny, taxonomy, and evolution of *Geosiphon pyriformis* and arbuscular mycorrhizal fungi. In *Diversity and Integration in Mycorrhizas* (pp. 75-83). Springer, Dordrecht.

Shi W, Zhang Y, Chen S, Polle A, Rennenberg H and Luo ZB (2019) Physiological and molecular mechanisms of heavy metal accumulation in nonmycorrhizal versus mycorrhizal plants. *Plant, Cell & Environment*, 42(4): 1087-1103.

Taylor DL and Sinsabaugh RL (2015) The soil fungi: occurrence, phylogeny, and ecology. *Soil Microbiology, Ecology, and Biochemistry*, 77-109.

Van der Heijden MG, Martin FM, Selosse MA and Sanders IR (2015) Mycorrhizal ecology and evolution: the past, the present, and the future. *New Phytologist*, 205(4): 1406-1423.

Vandenkoornhuyse P, Quaiser A, Duhamel M, Le Van A and Dufresne A (2015) The importance of the microbiome of the plant holobiont. *New Phytologist*, 206(4): 1196-1206.

Verma R, Tapwal A, Kumar D, Parkash V and Puri S (2019) Vesicular arbuscular mycorrhizal diversity in some important ethnomedicinal plants of Western Himalaya. *Med. Plants*, 11: 279-285.

Waqas MA, Kaya C, Riaz A, Farooq M, Nawaz I, Wilkes A and Li Y (2019) Potential mechanisms of abiotic stress tolerance in crop plants induced by thiourea. *Frontiers in Plant Science*, 10: 1336.

Wipf D, Krajinski F, Van Tuinen D, Recorbet G and Courty PE (2019) Trading on the arbuscular mycorrhiza market: from arbuscules to common mycorrhizal networks. *New Phytologist*, 223(3): 1127-1142.

Wright SF and Upadhyaya A (1996) Extraction of an abundant and unusual protein from soil and comparison with hyphal protein of arbuscular mycorrhizal fungi. *Soil Science*, 161(9): 575-586.

Wright SF and Upadhyaya A (1998). A survey of soils for aggregate stability and glomalin, a glycoprotein produced by hyphae of arbuscular mycorrhizal fungi. *Plant and Soil*, 198(1): 97-107.

Wu N, Li Z, Liu H and Tang M (2015) Influence of arbuscular mycorrhiza on photosynthesis and water status of *Populus cathayana* Rehder males and females under salt stress. *Acta Physiologiae Plantarum*, 37(9): 1-14.

Wu ZP, Wu WD, Zhou SL and Wu SH (2016) Mycorrhizal inoculation affects Pb and Cd accumulation and translocation in Pakchoi (*Brassica chinensis* L.). *Pedosphere*, 26(1): 13-26.

Yang Y, Han X, Liang Y, Ghosh A, Chen J and Tang M (2015) The combined effects of arbuscular mycorrhizal fungi (AMF) and lead (Pb) stress on Pb accumulation, plant growth parameters, photosynthesis, and antioxidant enzymes in *Robinia pseudoacacia* L. *PloS One*, 10(12): e0145726.

Yang Y, Liang Y, Han X, Chiu TY, Ghosh A, Chen H and Tang M (2016) The roles of arbuscular mycorrhizal fungi (AMF) in phytoremediation and tree-herb interactions in Pb contaminated soil. *Scientific Reports*, 6(1): 1-14.

Zhang XF, Hu ZH, Yan TX, Lu RR, Peng CL, Li SS and Jing YX (2019) Arbuscular mycorrhizal fungi alleviate Cd phytotoxicity by altering Cd subcellular distribution and chemical forms in Zea mays. *Ecotoxicology and Environmental Safety*, 171: 352-360.

Chapter 6

Occurrence and Role of Arbuscular Mycorrhizal (AM) Fungi in Mangrove Ecosystems

Sankrita Gaonkar*
Government College of Arts,
Science and Commerce, Quepem-Goa, India

Abstract

In recent years, the occurrence and ecological role of arbuscular mycorrhizal (AM) fungi in mangrove ecosystems have received increasing attention. Mangroves have great economic and ecological significance as they live in a specific type of habitat and hence are considered as one of the most important ecosystems.

In this chapter, the AM fungal status in mangroves of different regions of the world has been summarized. The AM fungi play a crucial role in the ecological development of the mangrove habitat. However, several factors such as soil flooding and salinity affect the survival of AM fungi in mangrove habitats. Based on the published literature, AM fungal species belonging to the family Glomeraceae are found to be dominant in the mangrove ecosystem followed by Acaulosporaceae species.

Keywords: arbuscular mycorrhiza, diversity, flooding, mangrove, salinity

---

* Corresponding Author's Email: sankrita002@gmail.com.

In: New Research on Mycorrhizal Fungus
Editors: Qiang-Sheng Wu, Ying-Ning Zou, Yue-Jun He et al.
ISBN: 979-8-88697-637-3

1. Introduction

Arbuscular mycorrhizal (AM) fungi are one of the most abundant groups of soil microbes (Parniske, 2008). They are obligate symbionts belonging to the phylum Glomeromycota forming a symbiotic relationship with around 90% of terrestrial plants (Smith and Read, 2008). These fungi assist plants in enhanced uptake of nutrients especially phosphorus (P) and in return receive carbon (C) from their host (Smith and Read, 2008).

Mangroves are one of the most important ecosystems found around the world having threats in their continuous existence (Wang et al., 2018). It is a special type of ecosystem occurring in intertidal, saline areas of tropical and sub-tropical regions. These plants include the plants that accomplish several vital socio-ecological functions by providing various valuable natural resources (Nagelkerken et al., 2008). Mangrove plants are divided into two categories: true mangroves and associate mangroves. True mangroves occur in the intertidal zones, whereas associate mangroves occur on landward fringes of the mangrove ecosystem (Wu et al., 2008).

According to the results of Šraj-Kržič et al., (2006), AM fungi play a marginal role in wetland habitats because of anoxic conditions which incline towards reduction of AM fungal colonization and sporulation. Nevertheless, there are a couple of studies that evidenced the occurrence and role of AM fungi in these habitats (Table 1) (D'souza and Rodrigues, 2013a,b; Wang et al., 2010, 2011, 2014; Gaonkar and Rodrigues, 2020, 2021). Earlier molecular studies by Wang (2011) indicated that flooded conditions inhibited the AM fungal and mangrove plant association. Soil salinity could be the next factor to eliminate AM fungal occurrence in mangrove habitats (Juniper and Abbott, 2006). However, Wang et al. (2011) implied the acquisition of oxygen by AM fungi from mangrove plant root aerenchyma under flooding conditions.

2.1. Composition and Diversity of Mangrove Plant Community

In an environment having a great influence of biotic and abiotic stresses, mycorrhizal plants can flourish better than non-mycorrhizal plants (D'Souza, 2016). A single plant can be colonized by one or more than AM fungal species which can form common mycorrhizal networks (CMN) (Jakobsen and Hammer, 2015). These interactions can develop stability in weaker plants that could improve nutrient supply through CMN (Van der Heijden and Horton, 2009).

Table 1. Status of arbuscular mycorrhizal (AM) fungi in mangroves

Source	Mangrove site	Plant name(s)	AM fungal species	Root colonization (%)	Spore density (spores/100 g of soil)
Kothamasi et al., 2006	Nicobar Island, India	*Rhizophora mucronata, Rhizophora stylosa, Bruguiera gymnorrhiza, Sonneratia acida, Nypa fruticans*	*Rhizophagus fasciculatus, Glomus magnicaule*	0-17.6	0-2
Kumar and Ghose, 2008	Sundarban mangroves, West Bengal, India	*Acanthus ilicifolius, Phoenix paludosa, Avicennia alba, A. marina, A. officinalis, Excoecaria agallocha, Xylocarpus granatum, X. mekongensis, Aegiceras corniculatum, Aegialitis rotundifolia, Porteresia coarctata, Bruguiera gymnorrhiza, B. parviflora, Ceriops decandra, C. tagal, Heritiera fomes*	*Acaulospora bireticulata, A. delicata, A. foveata, A. lacunosa, A. mellea, A. scrobiculata, Kuklospora spinosa, Gigaspora decipiens, Gi. margarita, Rhizophagus aggregatus, R. clarus, R fasciculatus, R. intraradices, Paraglomus albidum, P. lacteum, P. occultum, Funneliformis coronatus, F. geosporus, F. mosseae, F. verruculosus, Diversispora eburnea, Claroideoglomus etunicatum, C. drummondii, C. lamellosum, Gl. multicaule*	0.8 to 65.1	2.9-60.0
Wang et al., 2011	Zhuhai mangrove area, China	*Heritiera littoralis, Acrostichum aureum, Acanthus ilicifolius*	*A. foveata, R. intraradices, R. irregularis, F. mosseae, Sclerocystis sinuosa, Funneliformis dimorphicus, F. monosporus*	23.8-68.8	-

Table 1. (Continued)

Source	Mangrove site	Plant name(s)	AM fungal species	Root colonization (%)	Spore density (spores/100 g of soil)
Sridhar et al., 2011	Southwest coast, India	*Derris triflorum, Rhizophora mucronata, Canavalia cathartica, Acanthus ilicifolius, Cyperus malaccensis, Acrostichum aureum, Spinifex littoreus, Sonneratia alba*	*Dentiscutata nigra, D. erythropus, D. heterogama, F dimorphicus, Sieverdingia tortuosa, Acaulospora taiwania, A. denticulata, A. dilatata, A. rugosa, A. spinosa, R. aggregatus, R. fasciculatus, R. intraradices, Gigaspora gigantea, Rhizoglomus microaggregatum, P. occultum, Gi. margarita, Gi. albida, Racocetra gregaria, R. minuta, R. verrucose, R. weresubiae, Cetraspora pellucida, Septoglomus constrictum, Glomus reticulatum, Scutellospora calospora, S. dipurpurescens, S. projecturata, S. biornata, Entrophospora infrequens, Ambispora nicolsonii, Archaeospora undulata, Glomus australe, Gl. pallidum, F. mosseae*	7.8-69.0	67-430
Balachandran and Mishra, 2012	Western coast, Maharashtra, India	*Avicennia marina*	-	10-100	9-79

Source	Mangrove site	Plant name(s)	AM fungal species	Root colonization (%)	Spore density (spores/100 g of soil)
D'Souza and Rodrigues, 2013	Mangroves from Northern and Southern rivers of Goa, India	*Avicennia marina, Acanthus ilicifolius, Rhizophora apiculata, Ceriops tagal, Sonneratia caseolaris, Rhizophora mucronata, Avicennia officinalis, Excoecaria agallocha, Aegiceras corniculatum, Salvadora persica, Sonneratia alba, Acrostichum aureum, Derris heterophylla, Kandelia candel, Bruguiera cylindrica, Bruguiera gymnorrhiza*	*A. foveata, A. bireticulata, A. delicata, A. laevis, A. mellea, A. scrobiculata, A. spinosa, R. intraradices, R clarus, R. fasciculatus, R. aggregatus, F. mosseae, F. geosporus, Gl. multicaule, Gl. hyderabadensis, Gl. formosanum, Gl. nanolumen, S. constrictum, C. etunicatum, C. claroideum, R. gregaria, R. weresubiae, S. dipurpurescens, S. calospora, Gigaspora albida, Entrophospora infrequens, Sclerocystis taiwanensis, S. rubiformis*	6-77	0.08-3.2
Wang et al., 2014	Qi'Ao mangrove forest, China	*Heritiera littoralis, Pongamia pinnata, Cerbera manghas, Hibiscus tiliaceus*	*F. mosseae, Diversispora aurantia, Acaulospora longula, C. lamellosum, C. claroideum, C. etunicatum*	33.3-63.7	3.0-27.3
Hu et al., 2015	Mangrove forest, Southern China	*Aegiceras corniculatum, Acanthus ilicifolius*	-	0.3-36.5	54-206.8
Gopinathan et al., 2017	Muthupet mangrove area, Tamil Nadu, India	*Avicennia marina*	*A. bireticulata, A. delicata, A. spinosa, R. aggregatus, F. mossae, F. geosporus, Glomus microcarpum, S. constrictum, C. claroideum, S. rubiformis, D. heterogama, Gi. albida, Gi. margarita*	22.0-40.4	140-216

Table 1. (Continued)

Source	Mangrove site	Plant name(s)	AM fungal species	Root colonization (%)	Spore density (spores/100 g of soil)
da Silva et al., 2017	Ilha da Restinga, Brazil	-	*A. foveata, A. mellea, A. morrowiae, A. scrobiculata, Ambispora appendicula, Racocetra tropicana, R. coralloidea, Gi. margarita, Paradenstiscutata maritima, Cetraspora gilmorei, C. pellucida, Funneliformis halonatus, Glomus ambisporum, Gl. brohultii, G. glomerulatum, G. macrocarpum, G. microcarpum, S. sinuosa*	1.4-73.9	-
Fabián et al., 2018	State Reserve Ciénegas Manglares, Yucatan, Mexico	*Acoelorraphe wrightii*	*F. geosporus, F. verruculosus, R. clarus, R. intraradices, R. fasciculatus, S. sinuosa, S. rubiformis, Gl. microaggregatum, Gl. ambisporum, Gl. microcarpum, S. constrictum, C. claroideum, C. etunicatum, D. eburnea, S. heterogama, Gi. decipiens, A. morrowiae, A. scrobiculata, A. kentinensis, A. colombiana, P. occultum, E. infrequens*	25-60	126-211

Source	Mangrove site	Plant name(s)	AM fungal species	Root colonization (%)	Spore density (spores/100 g of soil)
Gaonkar & Rodrigues, 2020	Chorao Island, Goa	*Aegiceras corniculatum, Avicennia officinalis, Avicennia marina, Bruguiera cylindrica, Ceriops tagal, Excoecaria agallocha, Kandelia candel, Rhizophora apiculata, Rhizophora mucronata, Sonneratia alba, Sonneratia caseolaris, Acanthus ilicifolius, Acrostichum aureum, Clerodendrum inerme, Derris heterophylla, Pongamia pinnata, Thespesia populnea*	*A. dilatata, A. foveata, A. mellea, A. scrobiculata, A. undulata, A. denticulata, A. laevis, A. bireticulata, A. delicata, A. nicolsonii, A. rehmii, A. spinosa, Archaeospora myriocarpa, C. etunicatum*	20.0-97.5	20-138
Ramírez-Viga et al., 2020	Ría Lagartos Biosphere Reserve, Yucatán, Mexico	*Conocarpus erectus*	*Acaulospora scrobiculata, Funneliformis geosporus, F. coronatum, Claroideoglomus etunicatum, Scutellospora erythropus, Sclerocystis microcarpus, Acaulospora foveata, A. myriocarpa, A. appendicula, P. lacteum, G. microaggregatum, S. rubiformis*	40.7-71.0	7.6-15.2

Table 1. (Continued)

Source	Mangrove site	Plant name(s)	AM fungal species	Root colonization (%)	Spore density (spores/100 g of soil)
Gaonkar & Rodrigues, 2021	Pichavaram Forest, Tamil Nadu	*Aegiceras corniculatum, Avicennia marina, Avicennia officinalis, Brugueira cylindrica, Ceriops decandra, Excoecaria agallocha, Lumnitzera racemosa, Rhizophora apiculata, Rhizophora mucronata, Arthrocnemum indicum, Clerodendrum inerme, Salicornia brachiate, Ipomoea pes-caprae, Phoenix paludosa, Salvadora persica, Sesuvium portulacastrum, Suaeda monoica, Suaeda maritima*	*A. dilatata, A. foveata, A. nicolsonii, A. polonica, A. rehmii, A. scrobiculata, A. spinosa, A. undulata, Entrophospora* sp., *F. geosporus, F. mosseae, Gi. albida, Gi. decipiens, Gi. margarita, Gl. microcarpum, R. aggregatus, R. fasciculatus, R. intraradices, R. irregularis, Sclerocystis pachycaulis, S. rubiformis*	22.0-93.6	8-270
Akaji et al., 2021	Iriomote Island, Japan	*Rhizophora stylosa, Avicennia marina, Sonneratia alba, Bruguiera gymnorhiza, Baringtonia racemose, Heritiera littoralis, Pongamia pinnata*	*Acaulospora* sp., *Claroideoglomus* sp., *Diversispora* sp., *Glomus* sp., *Paraglomus* sp.	0-22	-
Martinez-Hernández, 2021	Tonala River, Mexico	*Laguncularia racemosa, Avicennia germinans*	*C. claroideum. Diversispora aurantium*	76	
Deepika and Kothamasi, 2021	Aghanashini and Gangavai River, India	*A. ilicifolius, A. officinalis, E. agallocha, R. apiculata, S. alba*	*Acaulospora* sp., *Entrophspora* sp., *Glomus* sp.	-	-

In a study of Wang et al. (2010), AM fungi could influence the composition of the mangrove plant community. Further, Daleo et al. (2008) suggested that the AM fungi affected the salinity zonation in salt marshes. Plant diversity was also influenced by AM fungi due to the facilitation of species competition that affected evenness in the plant population (Zhouying et al., 2016).

2. Roles of AM Fungi in Mangroves

AM fungi are reported to be abundant in wetlands, especially the mangrove ecosystem which can be comparable to that of terrestrial habitats (Wang et al., 2010; Deepika and Kothamasi, 2021).

Several biotic and abiotic environmental conditions such as inundation, edaphic factors, soil microflora, plant diversity, and litter decomposition control the availability of nutrients in the mangrove ecosystem (Wang et al., 2011; Tuheteru and Wu, 2017). The soil particles adsorb the insoluble form of P, while its soluble form is constantly used by the growing plants, resulting in the formation of a phosphate-free zone in the rhizosphere (Roy-Boldoc, 2011). Nonetheless, soil microbial flora like AM fungi are well known to enhance nutrient uptake especially P (Aggarwal et al., 2012; Willis et al., 2013). AM fungi do this by forming a huge extraradical hyphal network that can penetrate the phosphate-free zone and lengthens host P absorption (Xie et al., 2014).

2.1. Nutrition in Mangrove Plants

As discussed earlier, AM fungal association with its host can enhance nutrient uptake. This includes vital plant nutrients like P, N, Cu, Zn, S, Mg, Mn, Fe, etc. (Leigh et al., 2008).

Yu et al., (2021) conducted a comprehensive study on the effect of the association between AM fungi and diazotrophs on mangrove habitats. Their results showed that the positive interaction between AM fungi and diazotrophs stimulated N fixation efficacy in mangrove plants. AM fungal species are capable of regulating N-linked enzyme activities (Nuccio et al., 2013). Xie et al. (2014) evaluated the dual effect of both AM fungi and P supply on plant growth and nutrition of *Kandelia obovata* plants. They found that inoculation with AM fungi and 30 mg/kg supply of KH_2PO_4 boosted root strength, sustained activities of soil acid phosphatase (ACP) and alkaline phosphatase (ALP), and enhanced N and P uptake, thus leading to increased biomass of *K. obovata* plants. In a greenhouse experiment performed by Wang et al. (2010), the effect of AM fungal inoculation on the growth and nutrient uptake of *Sonneratia apetata* plants was estimated. This experiment indicated that the AM fungal inoculation increased plant height and biomass, along with enhance uptake of N, P, and K.

2.2. Stress Tolerance

AM fungi provide an ecological competitive advantage to their host allowing better growth and survival under unfavorable environmental conditions such as low temperature, acid stress, soil moisture, soil salinity, etc. (Mohammadi et al., 2011). AM fungi naturally occur in saline habitats inspite of having inhibitory effects on their survival (Aggarwal et al., 2012). It is well demonstrated that AM fungi can alleviate salinity stress in plants by undergoing several mechanisms such as improved nutrient acquisition (Evelin et al., 2012) and by regulating the physiological functions of their host plants (Latef and Miransari, 2014; Chang et al., 2018). It was suggested by Porras-Soriano et al., (2009) that the efficiency of alleviating salt stress in plants depended on the type of AM fungal species.

3. Factors Affecting AM Fungi in the Mangrove Ecosystem

3.1. Flooding

Although it is well understood that AM fungi assist in enhanced nutrient uptake, their efficacy of nutrient mobilization decreases in the waterlogged conditions of the mangrove ecosystem. Wang et al. (2011) reported reduced AM fungal diversity and colonization under intensive flooding. AM fungi are aerobic microbes, and this could be attributed to the fact that they could survive in aerobic regions of mangrove plant roots. Nevertheless, AM fungi are inhibited by the oxygen-deprived zone outside the rhizosphere in recurrently inundated regions of the mangrove ecosystem. Hence, AM fungal germination and colonization may get repressed in such zones (Wang et al., 2011).

3.2. Salinity

Soil salinity can negatively affect AM fungal spore germination, root colonization, and hyphal growth (Latef and Miransari, 2014). The colonization rate of AM fungi reduces with increasing concentration of NaCl and same might delay spore germination (Hajiboland, 2013). The presence of numerous salts in soil inhibits hyphal growth thus decreasing the spread of the hyphal network (Latef and Chaoxing, 2014). Juniper and Abbott (2006) reported that the salts like $NaNO_3$ and Na_2SO_4 that have similar osmotic potentials affect spore germination differently. In saline soils, spores cannot get hydrated which is important for its germination and consequently delays the process (Evelin et al., 2009). Delvian and Rambey (2019) observed a decrease in the colonization by *Gigaspora margarita* at a high concentration of NaCl, while *Glomus etunicatum* showed a significant rate of colonization even at 10,000 μg/g NaCl.

Peng et al., (2010) in their study observed that the application of NaCl decreased colonization by two AM fungal species, *Glomus mosseae* and *Claroideoglomus claroideum*, whereas it did not affect the colonization by *Rhizophagus intraradices*. Gupta et al. (2016) reported the occurrence of *Glomus* and *Scutellospora* across low to high saline zones, while the species of *Acaulospora* inhabited low salinity areas. Using meta-analyses, Bothe (2012) examined AM fungal studies on salinity stress and suggested that the germination and hyphal growth of AM fungal spores were affected by salt stress.

Conclusion and Outlook

Moisture content in the soil is the dominant factor in mangroves as it is the determining factor of the soil attributes and the type of microflora that live within the rhizosphere. In the case of terrestrial ecosystems, AM fungi show the high density and diversity in the mangrove ecosystems. Their vital roles in the composition, diversity, ecological succession, nutrition, and productivity of mangrove plants have been proven in various studies. However, AM fungal roles in increasing photosynthesis, secondary metabolites, providing disease resistance to their mangrove host, and hydrocarbon degradation in mangrove habitats have not been suitably investigated, which requires a deeper focus on the mangrove ecosystem, especially the functions performed by the mycorrhizal fungi within the mangrove ecosystem. This review offers scope for future studies which could be carried out on above mentioned unexplored areas of AM fungal applications in the mangrove ecosystems.

References

Aggarwal A., Kadian N., Karishma N., Tanwar A., Gupta K. K. (2012) Arbuscular mycorrhizal symbiosis and alleviation of salinity stress. *Journal of Applied and Natural Science*, 4:144-155.

Akaji Y., Inoue T., Taniguchi T., Baba S. (2021) Arbuscular mycorrhizal fungal communities of a mangrove forest along a salinity gradient on Iriomote Island. *Plant and Soil*, 472:145-159.

Balachandran S., Mishra S. (2012) Assessment of arbuscular mycorrhizal fungi (AM fungi) and glomalin in the rhizosphere of heavy metal polluted mangrove forest. *International Journal of Environmental Science*, 4:392-401.

Bothe H. (2012) Arbuscular mycorrhiza and salt tolerance of plants. *Symbiosis*, 58:7-16.

Chang W., Sui X., Fan X. X., Jia T. T., Song F. Q. (2018) Arbuscular mycorrhizal symbiosis modulates antioxidant response and ion distribution in salt-stressed *Elaeagnus angustifolia* seedlings. *Frontiers in Microbiology*, 9:652.

D'Souza J. (2016) Arbuscular mycorrhizal diversity from mangroves: A Review. In: Pagano MC (ed) *Recent Advances on Mycorrhizal Fungi, Fungal Biology*. Springer International Publishing Switzerland, pp. 109-116.

D'Souza J., Rodrigues B. F. (2013) Biodiversity of arbuscular mycorrhizal (AM) fungi in mangroves of Goa in West India. *Journal of Forest Research*, 24:515–523.

da Silva I. R., da Silva D. K. A., de Souza F. A., Oehl F., Maia L. C. (2017) Changes in arbuscular mycorrhizal fungal communities along a river delta island in Northeastern Brazil. *Acta Oecologica, 79*:8-17.

Daleo P., Alberti J., Canepuccia A., Escapa M., Fanjul E., Silliman B. R., Bertness M. D., Iribarne O. (2008) Mycorrhizal fungi determine salt-marsh plant zonation depending on nutrient supply. *Journal of Ecology*, 96:431-437.

Deepika S., Kothamasi D. (2021) Plant hosts may influence arbuscular mycorrhizal fungal community composition in mangrove estuaries. *Mycorrhiza*, 31:699-711.

Delvian, Rambey R. (2019) Effect of salinity on spore germination, hyphal length and root colonization of the arbuscular mycorrhizal fungi. *IOP Conf. Series: Earth and Environmental Science*, 260: 012124.

Evelin H., Giri B., Kapoor R. (2012) Contribution of *Glomus intraradices* inoculation to nutrient acquisition and mitigation of ionic imbalance in NaCl-stressed *Trigonella foenum-graecum*. *Mycorrhiza*, 22:203-217.

Evelin H., Kapoor R., Giri B. (2009) Arbuscular mycorrhizal fungi in alleviation of salt stress: a review. *Annal of Botany*, 104(7): 1263–1280.

Fabián D., Guadarrama P., Hernadez-cuevas L., Ramos-Zapata J. A. (2018) Arbuscular mycorrhizal fungi in a coastal wetland in Yucatan, Mexico. *Botanical Sciences*, 96: 24-34.

Gaonkar S., Rodrigues B. F. (2020) Diversity of arbuscular mycorrhizal (AM) fungi in mangroves of Chorao Island, Goa, India. *Wetlands Ecology and Management*, 28:765-778.

Gaonkar S., Rodrigues B. F. (2021) Arbuscular mycorrhizal fungal status in mangroves of Pichavaram Forest, Tamil Nadu, India. *Tropical Ecology*, 62:538-548.

Gopinathan M., Mahesh V., Durgadevi R. (2017) Occurrence and distribution of AM fungi associated in *Avicennia marina* (Forssk) Vierh in soils of south-east coastal area of Muthupet, India. *Journal of Global Biosciences*, 6:4741-4747.

Gupta N., Bihari K. M., Sengupta I. (2016) Diversity of arbuscular mycorrhizal fungi in different salinity of mangrove ecosystem of Odisha, India. Advances in Plants & *Agriculture Research*, 3:19-23.

Hajiboland R. (2013) Role of arbuscular mycorrhiza in amelioration of salinity. In: Ahmad P., Azooz M. M., Prasad M. N. V. (eds) *Salt Stress in Plants: Signalling, Omics and Adaptations*. Springer, New York, pp. 301-354.

Hu W., Wu Y., Xin G., Wang Y., Guo J., Peng X. (2015) Arbuscular mycorrhizal fungi and their influencing factors for *Aegiceras corniculatum* and *Acanthus ilicifolius* in southern China. *Pakistan Journal of Botany*, 47:158-1586.

Jakobsen I., Hammer E. C. (2015) Nutrient dynamics in arbuscular mycorrhizal networks. In: Horton T. R. (ed) *Mycorrhizal Networks*. Springer Dordrecht Heidelberg London New York, pp. 91-131.

Juniper S., Abbott L. K. (2006) Soil salinity delays germination and limits growth of hyphae from propagules of arbuscular mycorrhizal fungi. *Mycorrhiza*, 16:371-379.

Kothamasi D., Kothamasi S., Bhattacharyya A., Kuhad R. C., Babu C. R. (2006) Arbuscular mycorrhizae and phosphate solubilising bacteria of the rhizosphere of the mangrove ecosystem of Great Nicobar Island, India. *Biology and Fertility of Soils*, 42:358-361.

Kumar T., Ghose M. (2008) Status of arbuscular mycorrhizal fungi (AMF) in the Sundarbans of India in relation to tidal inundation and chemical properties of soil. *Wetlands Ecology and Management*, 16:471-483.

Latef A. A., Chaoxing H. (2014) Does the inoculation with *Glomus mosseae* improve salt tolerance in pepper plants? *Journal of Plant Growth Regulation*, 33:644-653.

Latef A. A., Miransari M. (2014) The role of arbuscular mycorrhizal fungi in alleviation of salt stress. In: Miransari M. (ed) *Use of Microbes for the Alleviation of Soil Stresses*. Springer Science Business Media, New York, pp. 23-38.

Leigh J., Hodge A., Fitter A. H. (2008) Arbuscular mycorrhizal fungi can transfer substantial amounts of nitrogen to their host plant from organic material. *New Phytology*, 181:199-207.

Martínez-Hernández I. A., Rivera-Cruz M. C., Carballar-Hernández S., Trujillo-Narcía A., Ortíz-García C. F., Hernández-Galvez G. and Alarcón A. (2021) Arbuscular mycorrhizal colonization in a mangrove forest exposed to weathering oil for half a century. *Water Air and Soil Pollution*, 232:34.

Mohammadi K., Shiva Khalesro, Sohrabi Y., Heidari G. (2011) A review: Beneficial effects of the mycorrhizal fungi for plant growth. *Journal of Applied Environmental and Biological Sciences*, 1:310–319.

Nagelkerken I., Blaber S. J. M., Bouillon S., Green P., Haywood M., Kirton L. G., Meynecke J. O., Pawlik P. H. M., Sasekumar A., Somerfield P. J. (2008) The habitat function of mangroves for terrestrial and marine fauna: a review. *Aquatic Botany*, 89:155-185.

Nuccio E. E., Hodge A., Pett-Ridge J., Herman D. J., Weber P. K., Firestone M. K. (2013) An arbuscular mycorrhizal fungus significantly modifies the soil bacterial community and nitrogen cycling during litter decomposition. *Environmental Mirobiology*, 15:1870-1881.

Parniske M. (2008) Arbuscular mycorrhiza: the mother of plant root endosymbioses. *Nature Reviews Microbiology*, 6:763–775.

Peng J., Li Y., Shi P., Chen X., Lin H., Zhao B. (2010) The differential behavior of arbuscular mycorrhizal fungi in interaction with *Astragalus sinicus* L. under salt stress. *Mycorrhiza*, 21:27–33.

Porras-Soriano A., Soriano-Martín M. L., Porras-Piedra A., Azcón R. (2009) Arbuscular mycorrhizal fungi increased growth, nutrient uptake and tolerance to salinity in olive trees under nursery conditions. *Journal of Plant Physiology*, 166:1350-1359.

Ramírez-Viga T. K., Ramos-Zapata J. A., Cantón C. G., Hernández-Cuevas L., Guadarrama-Chávez P. (2020) Arbuscular mycorrhizal association in *Conocarpus erectus* (Combretaceae) in mangroves from Yucatán, Mexico. *Botanical Sciences*, 98: 66-75.

Roy-Bolduc A., Hijri M. (2011) The use of mycorrhizae to enhance phosphorus uptake: a way out the phosphorus crisis. *Journal of Biofertilizers and Biopesticides*, 2:1-5.

Smith S. E., Read D. J. (2008) *Mycorrhizal Symbiosis.* Cambridge: Academic Press.

Šraj-Kržič N., Pongrac P., Klemenc M., Kladnik A., Regvar M., Gaberscik A. (2006) Mycorrhizal colonisation in plants from intermittent aquatic habitats. *Aquatic Botany*, 85:331-336.

Sridhar K. R., Roy S., Sudheep N. M. (2011) Assemblage and diversity of arbuscular mycorrhizal fungi in mangrove plant species of the southwest coast of India. In: Metras J. (ed) *Mangroves Ecology, Biology and Taxonomy*. Nova Science Publishers Inc, Hauppage, pp. 257-274.

Tuheteru F. D., Wu Q. S. (2017) Arbuscular mycorrhizal fungi and tolerance of waterlogging stress in plants. In: Wu QS (ed) *Arbuscular Mycorrhizas and Stress Tolerance of Plants*. Springer, Singapore, pp. 43-66.

Van Der Heijden M. G., Horton T. R. (2009) Socialism in soil? The importance of mycorrhizal fungal networks for facilitation in natural ecosystems. *Journal of Ecology*, 97:1139-1150.

Wang Y., Huang Y., Qiu Q., Xin G., Yang Z., Shi S. (2011) Flooding greatly affects the diversity of arbuscular mycorrhizal fungi communities in the roots of wetland plants. *PLoS ONE*, 6: e24512.

Wang Y., Li T., Li Y., Qiu Q., Li S., Xin G. (2015) Distribution of arbuscular mycorrhizal fungi in four semi-mangrove plant communities. *Annals of Microbiology*, 65:603-610.

Wang Y., Qui Q., Yang Z., Hu Z., Tam N. F., Xin G. (2010) Arbuscular mycorrhizal fungi in two mangroves in South China. *Plant and Soil*, 331:181-191.

Willis A., Rodrigues B. F., Harris P. J. C. (2013) The ecology of arbuscular mycorrhizal fungi. *Critical Reviews in Plant Sciences*, 32:1-20.

Wu J., Xiao Q., Xu J., Li M. Y., Pan J. Y., Yang M. H. (2008) Natural products from true mangrove flora: source, chemistry and bioactivities. *Natural Product Reports*, 25:955-981.

Xie X., Weng B., Cai B., Dong Y., Yan C. (2014) Effects of arbuscular mycorrhizal inoculation and phosphorus supply on the growth and nutrient uptake of *Kandelia obovata* (Sheue, Liu & Yong) seedlings in autoclaved soil. *Applied Soil Ecology*, 75:162-171.

Yu H., Liu X., Yang C., Peng Y., Yu X., Gu H., Zheng X., Wang C., Xiao F., Shu L., He Z., Wu B., Yan Q. (2021) Co-symbiosis of arbuscular mycorrhizal fungi (AMF) and diazotrophs promote biological nitrogen fixation in mangrove ecosystems. *Soil Biology and Biochemistry*, 161:108382.

Zhouying X., Yihui B., Yinghe J., Xiangling Z., Xiaoying L. (2016) Arbuscular mycorrhizal fungi in wetland habitats and their application in constructed wetland: A review. *Pedosphere*, 26: 592-617.

Chapter 7

In Vitro Mycorrhization as a Keystone of Successful In Vitro Pear Production

Mariem Lotfi
and Stefaan P.O. Werbrouck*
Laboratory for Applied In Vitro Plant Biotechnology,
Department of Plants and Crops,
Faculty of Bioscience Engineering,
Ghent University, Ghent, Belgium

Abstract

Pear trees rank third most important place of all fruit trees, after apple trees and grapevines. In this chapter, the mycelium donor plant (MDP) system was applied to pear, in which inoculation of pear plants with extraradical mycelium of *R. irregularis* from a donor plant enabled rapid and efficient *in vitro* colonisation. This system can be adapted to other plant species and allows the production of efficiently mycorrhized *in vitro* plants, which might better cope with the stress of adaptation to soil and climatic conditions under field conditions.

Keywords: arbuscular mycorrhizal fungi, mycelium donor plant (MDP) system, pear

1. Introduction

In temperate zones, pear trees (*Pyrus* spp.) occupy the third most important place of all fruit trees, after apple trees and grapevines. Due to their high level

* Corresponding Author's Email: stefaan.werbrouck@ugent.be.

In: New Research on Mycorrhizal Fungus
Editors: Qiang-Sheng Wu, Ying-Ning Zou, Yue-Jun He et al.
ISBN: 979-8-88697-637-3

of heterozygosity, they are reproduced by grafting. The choice of rootstock determines growth and development to a significant degree. Dwarfing rootstocks proved to be the key to a shortened juvenile period and improved fruit quality. In Europe, quinces are most commonly used as rootstock. However, these prove less suitable in the Mediterranean region, which is plagued by extreme drought and high salt stress. Therefore, *Pyrus communis* rootstocks are also being researched for application in countries such as Tunisia.

Local cultivars, with low cold requirements, are traditionally cultivated in coastal regions and present a phenotypic and genetic diversity at the origin of certain highly sought-after characteristics. European and American cultivars are cultivated more or less intensively in continental regions of the country with cold winters (Lotfi et al., 2019a). Despite the economic and nutritional importance of pear cultivation in Tunisia, it faces several abiotic and biotic constraints limiting its extension. The poor choice of varieties and climatic variations represent the main cause of the extinction of local cultivars and rootstocks, aggravated by the absence of well-defined standards in terms of crop management (Lotfi et al., 2019a). Especially since in 2012, the fire blight disease, caused by *Erwinia amylovora*, ravaged the cultures of *Pyrus communis* 'Alexandrine' and 'Williams' (Rhouma et al., 2014); heavily affecting our cultivars by eradicating more than 5500 ha of plantation out of a total of 8400 ha (Gaaliche et al., 2018).

Micropropagation plays an important role in the rapid propagation of disease-free, high-quality plant material. The application of this technology in pears for use as rootstock is still in its infancy. Although micropropagation is feasible (Lotfi et al., 2019a), rooting and acclimatization remain key steps in micropropagation. They are the tailpiece of the production process of pear rootstocks to meet the increased needs for pears on the national market.

Transferring *in vitro* plants to an *in vivo* environment necessitates careful manipulation to minimize transplant shock and to augment survival rates. This is the phase when a functional root system develops (Kapoor et al., 2008). In this context, Varma and Schüepp (1995) explained the limited survival and establishment rates of micropropagated plants during transplantation by the lack of associated natural microorganisms. According to Shishikura et al., (2021), arbuscular mycorrhizal fungi (AMF) can be added (i) *in vitro*, during the rooting stage; (ii) *in vivo*, at the beginning of the weaning stage, and (iii) *in vivo*, after the acclimatization stage, but prior to the beginning of the post-acclimatization phase in the greenhouse (the hardening off stage). It is worth noting that the most common mycorrhization method is to mix the plants with the substrate *in vivo* (stages ii and iii). This technique enhanced rooting, biomass production, and survival percentage in various woody species such as olive (Ouledali et al., 2018), and date palm (El Hilali et al., 2021). However, plant roots can also interact with mycorrhiza *in vitro* (stage i). It is a complex process that uses various hosts. This review discusses the current literature and explains the use of the so-called Mycelium Donor Plant (MDP) system in pears.

2. *In Vitro* Plant: Arbuscular Mycorrhizal Fungi Interactions

AMF plant inoculation in the *in vitro* phase was reported by potato (Voets et al., 2005) and strawberry (Cassells et al., 1996). It was particularly studied in banana (Koffi et al., 2009) and rubber tree (Sosa-Rodriguez et al., 2013). The reported benefits of introducing arbuscular mycorrhizal fungi (AMF) during the *in vitro* stage are several: (1) the plantlets are small and do not necessitate large amounts of inoculum; (2) new growing roots are suitable for an effective association before suberization; (3) there is no competition with other microbes during *in vitro* cultivation; (4) the defense mechanism developed *in vitro* endows plants with better resistance to biotic and abiotic pressure during transfer (Gallou et al., 2010).

The *in vitro* inoculation of Ri-T-DNA transformed roots has been investigated over the last decades (Bécard and Fortin, 1988), allowing AMF to be studied *in vitro*. Carrot roots in particular have been commonly associated with AMF (Declerck et al., 1998), but isolated hairy roots of *Cichorium intybus* and *Medicago truncatula* have similarly been used. AMF can also finish their life cycle on rooted *in vitro* plants. The advantage of such approaches is that they allow for detailed observations in a non-destructive way. Using these techniques, Declerck et al. (1996), Voets et al. (2006) and Declerck et al. (2001) investigated the developmental dynamics of the AMF, the architecture of the mycelium and the ontogeny of the spores. By upgrading this method with two compartments that allow the mycelium to escape from the donor roots and later colonize the receiver plant (Lalaymia and Declerck, 2020), it was also possible to examine the transport of elements to the root (de Boulois et al., 2005). This separation also allowed for the isolation and identification of AMF pure strains, which was in turn beneficial for molecular analyses (Koch et al., 2006).

In the 1990s, Elmeskaoui et al. (1995) and Cassells et al. (1996) were the first to develop tripartite (*Fragaria vesca* + AMF associated with carrot root) and bipartite (*Fragaria vesca* + AMF from isolated spores) *in vitro* culture systems for the colonization of micropropagated plantlets. These systems were employed to explore the effects of water stress (Hernández-Sebastià et al., 1999) and of light and carbon dioxide on mycorrhizae development (Louche-Tessandier et al., 1999). Lately, Voets et al. (2006) devised an autotrophic *in vitro* culture system that associated *Glomus intraradices* spores, produced by transformed carrot hairy roots, with potatoes. This system resulted in considerable root colonization and in the growth of thousands of spores. Koffi et al. (2009) used this system with banana plants and achieved a comparatively limited root colonization after eleven weeks. This was considerably long for an optimal colonization within the time lapse required for *in vitro* plant micropropagation.

Voets et al. (2009) developed the 'Mycelium Donor Plant' (MDP) in which micropropagated plants could be pre-colonized via an expanding extraradical mycelial network that was obtained from the mycorrhized roots of a donor plant such as alfalfa, and strawberry. Thereby, colonization occurred uniformly, and the required time duration was reduced to two weeks in *Solanum tuberosum* plants. Experiments on banana, rubber tree, and pear plants followed. Koffi and Declerck (2015) documented a considerable colonization rate of banana roots: nearly 40% of the roots comprised arbuscules, and in more than 20% root spores/vesicles were recorded within three weeks. When transferred to a fresh culture medium, colonized banana plants were able to reproduce the fungal life cycle, yielding thousands of spores during 3 weeks. As documented by Sosa-Rodriguez et al. (2013), arbuscular growth at high carbon dioxide concentration was recorded at 21.5% and spore growth at 19.1% of the roots. Lotfi et al. (2019) also reported that, in the case of pear, after 1 month of plantlet contact with the MDP system, about 30% of the root system was colonized, with 29% arbuscules and 30% spores/vesicles.

3. Application of the Mycelium Donor Plant (MDP) System on Pears

3.1. Overview of the Biological Material

The starting material consists of a culture of hairy roots of Ri T-DNA transformed *Cichorium intybus* associated with *Rhizophagus irregularis*.

Medicago truncatula is used as the mycelium donor plant, while the recipient plants are rooted *in vitro* plants of *Pyrus communis* plants.

3.2. *In Vitro* Culture of Pears

Non-woody cuttings of about 20 cm in length, previously stripped of their leaves, are washed thoroughly. Pieces with 3 to 4 internodes are incubated in a mixture of 0.2% ascorbic acid and 0.1% citric acid, to prevent oxidation by phenols released from the explants, and then shaken in 70% ethanol for 1 minute. This is followed by immersion for 2 min in a 0.6% benomyl solution. Finally, the explants are sterilized externally in a 0.4% mercuric chloride solution for 20 minutes, to which 2 drops of Tween-20 detergent are added. Finally, after they are rinsed three times with sterile distilled water, top nodes of 3 cm are transferred to tubes and later micropropagated in a modified MS medium with half a concentration of NH_4NO_3 and KNO_3, 30 g/L sucrose, 0.5 g/L calcium gluconate, 5 μM *meta*-topolin riboside (*m*TR) and 7 g/L agar–agar (pH 5.8). The cultures are kept at 25 ± 1 °C under a 16/8 h light/dark photoperiod under cool fluorescent light. For *in vitro* rooting, shoots are placed in ECO2BOX containers with filter (XL = No. 20) (Microbox™, Belgium) containing 150 ml of half-strength MS medium, enriched with 5 μM *trans*-cinnamic acid (*t*-CA) which is added after autoclaving (Lotfi et al., 2019a).

3.3. Mycelium Donor Plant (MDP) System Setup

As in the case of their maintenance, plugs of the hairy root culture are transferred to Petri dishes (Ø 90 mm) with 35 ml Strullu-Romand medium (MSR) (Declerck et al., 1998). The Petri dishes are incubated at 27 °C in the dark in an inverted position until spores are formed (Figure 1A).

With the aid of heated forceps, a 2-mm-diameter hole is punched in the side of each of the 90-mm-diameter Petri dishes used. A volume of 20-ml modified, vitamin and sucrose-free MSR medium is added to each Petri dish. Before autoclaving, the medium is set at pH 5.4.

Seeds of *Medicago truncatula* are surface sterilized by immersion in sodium hypochlorite (8% active chloride) for 15 min, rinsed three times with sterile distilled water, germinated in Petri dish (Ø 90 mm) with 35 ml of MSR medium lacking sucrose and vitamins.

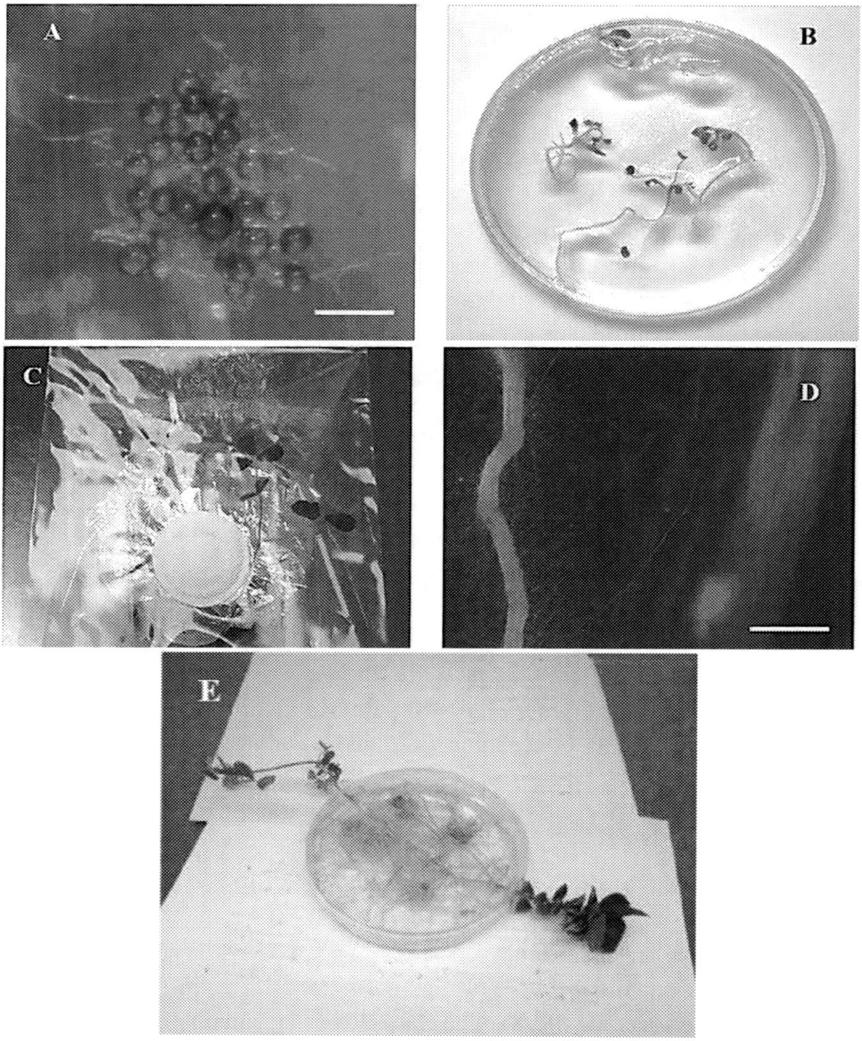

Figure 1. *In vitro* colonization method of *Pyrus communis* plantlets. (A) *M. truncatula* 'Salsa' seedling growing in MSR medium; (B) *Agrobacterium rhizogenes* transformed chicory root hair grown on MSR medium; (C) plastic bag with Petri dishs wrapped in aluminum foil to keep the AMF and the chicory hairy roots growing in darkness, while the shoots of *M. truncatula* can still grow under light conditions; (D) *R. irregularis* culture associated with chicory root and spore development; (E) *In vitro M. truncatula* seedlings and rooted pear plant nine weeks after the setup of the MDP system. Figure 1A and 1D: magnification ratio 0.8 x (Bar: 100μm).

Well-developed *Medicago truncatula in vitro* seedlings (7-day-old) showing a well-developed primary root are transferred to the Petri dishes (Figure 1B). The root system is placed in MSR medium in the Petri dish, and the stem is arranged to emerge outwards through the previously punctured hole. A gel plug containing about 50 *R. irregularis* spores is positioned near the roots. Immediately after transferring the seedlings, the Petri dish is closed and taped with a Parafilm™ strip. The hole around the seedling stem is sealed in the same way with sterile silicone glue to prevent contamination. Then the Petri dishes are wrapped in aluminium foil in a plastic bag, to keep the AMF and the *M. truncatula* roots in the dark, while the stem continues to grow in the light (Figure 1C). The Petri dishes are placed in a growth chamber under controlled conditions (25°C, with a 16 h day-1 photoperiod and 50 μmol.m$^{-2.s-1}$ photosynthetic photon flux). Starting at week 2, a 10 ml volume of vitamin and sucrose free MSR medium, cooled to 40°C in a water bath, is weekly poured into the Petri plates. Five weeks later, the extraradical mycelium begins to colonize the Petri plate, and in a period of 4 weeks covers the entire surface of the medium (Figure 1D).

3.4. Addition of the Receiver Plants

Nine weeks after the set-up, a rooted pear plant is transferred into each MDP system (Figure 1E). The Parafilm strips are withdrawn and the MDP systems opened under a laminar flow hood. A 2 mm diameter hole is perforated in the Petri plate with heated forceps. The holes at the top and bottom of the Petri dish must be perfectly aligned. After puncturing the hole, a well-rooted pear plant is inserted into each notch of the Petri dish.

The roots of the pear plant are carefully placed on the surface of the medium in direct contact with the AMF mycelium (Lotfi et al., 2019b). The Petri dishes are then closed and sealed with a Parafilm strip. As with the *M. truncatula* stems, the opening around the pear stems is plugged with sterile silicone gel. The MDP systems are then incubated as described above.

4. Acclimatization

After 30 days, the pear plants are transferred to pots with coco peat and perlite without adding fertilizer to acclimatize in a mist-growth chamber with approximately 90% relative humidity for 2 weeks at 26 ± 1°C under a 16-hour photoperiod of white fluorescent light. Then the plants are transferred to polyethylene pots, with the same substrate composition, and kept in a growth room at 21°C and 60% RH. Mycorrhizal colonization had shown in roots (Figure 2A-C). Figure 3 illustrates the better root development of the mycorrhized (right) compared to the non-mycorrhized (left) roots after one month of acclimatization, which promises improved performance in field conditions.

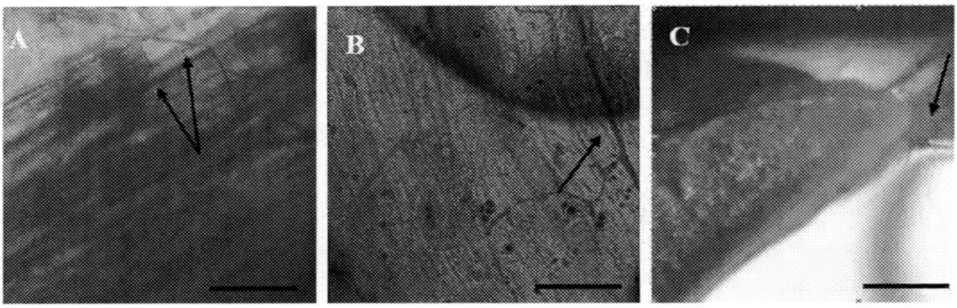

Figure 2. The root system of *Pyrus communis* colonized by *R. irregularis*.
[A]: Presence of spore and mycelium in root sections mycorrhized by AMF (arrow).
[B]: hyphae detected inside the root (arrow). [C]: Appressorium (arrow).
Magnification ratio 0.8 x (Bar: 100μm).

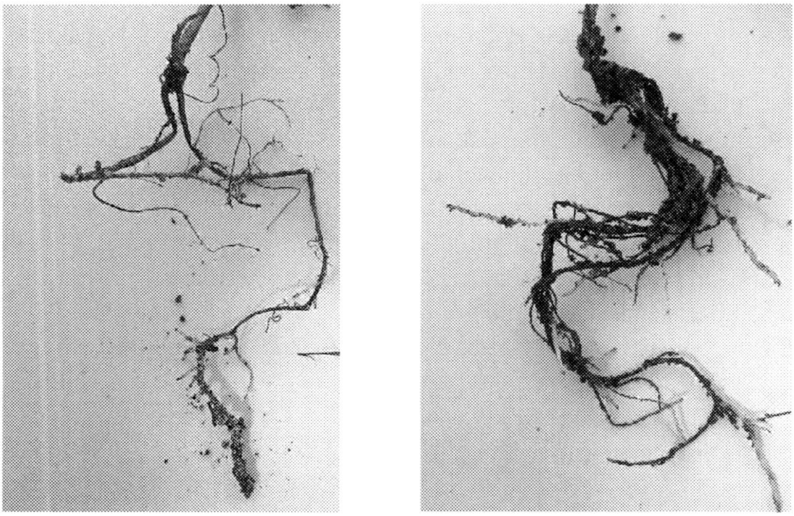

Figure 3. Colonized (right) and non-colonized (left) root of pear after one month of acclimatization.

Conclusion and Outlook

Inoculation of pear plants with extraradical mycelium of *Rhizophagus irregularis* from a donor plant enables rapid and efficient *in vitro* colonization. This system can be adapted to other plant species and will allow the production of efficiently mycorrhized *in vitro* plants, which might better cope with the stress of adaptation to soil and climatic conditions under field conditions.

References

Bécard G, Fortin J A (1988) Early events of vesicular–arbuscular mycorrhiza formation on Ri T-DNA transformed roots. *New Phytologist*, 108(2), 211-218.

Cassells A C, Mark GL, Periappuram C (1996) Establishment of arbuscular mycorrhizal fungi in autotrophic strawberry cultures *in vitro*. Comparison with inoculation of microplants *in vivo*. *Agronomie*, 16, 625-632.

de Boulois H D, Delvaux B, Declerck S (2005) Effects of arbuscular mycorrhizal fungi on the root uptake and translocation of radiocaesium. *Environmental Pollution,* 134(3), 515-524.

Declerck S, D'or D, Cranenbrouck S, Boulengé L E (2001) Modelling the sporulation dynamics of arbuscular mycorrhizal fungi in monoxenic culture. *Mycorrhiza,* 11(5), 225-230.

Declerck S, Strullu D G, Plenchette C (1996) *In vitro* massproduction of the arbuscular mycorrhizal fungus, *Glomus versiforme*, associated with Ri T-DNA transformed carrot roots. *Mycological Research, 100(10), 1237-1242.*

Declerck S, Strullu D G, Plenchette C (1998) Monoxenic culture of the intraradical forms of *Glomus* sp. isolated from a tropical ecosystem: a proposed methodology for germplasm collection. *Mycologia,* 90(4), 579- 585.

El Hilali R, Bouamri R, Crozilhac P, Calonne M, Symanczik S, Ouahmane L, Declerck S (2021) *In vitro* colonization of date palm plants by *Rhizophagus irregularis* during the rooting stage. *Symbiosis,* 84(1), 83-89.

Elmeskaoui A, Damont J P, Poulin M J, Piché Y, Desjardins Y (1995) A tripartite culture system for endomycorrhizal inoculation of micropropagated strawberry plantlets *in vitro*. *Mycorrhiza*, 5(5), 313-319.

Gaaliche B, Chehimi S, Dardouri S, Hajlaoui MR (2018) Health status of the pear tree following the establishment of fire blight in Northern Tunisia. *International Journal of Fruit Science*, 18(1), 85-98.

Gallou, A., De Jaeger, N., Cranenbrouck, S., & Declerck, S. (2010). Fast track *in vitro* mycorrhization of potato plantlets allow studies on gene expression dynamics. *Mycorrhiza*, 20(3), 201-207.

Hernández-Sebastià C, Piché Y, Desjardins Y (1999) Water relations of whole strawberry plantlets *in vitro* inoculated with *Glomus intraradices* in a tripartite culture system. *Plant Science*, 143(1), 81-91.

Kapoor R, Sharma D, Bhatnagar AK (2008) Arbuscular mycorrhizae in micropropagation systems and their potential applications. *Scientia Horticulturae*, 116(3), 227-239.

Koch A M, Croll D, Sanders I R (2006) Genetic variability in a population of arbuscular mycorrhizal fungi causes variation in plant growth. *Ecology Letters*, 9(2), 103-110.

Koffi M C, de la Providencia I E, Elsen A, Declerck S (2009) Development of an *in vitro* culture system adapted to banana mycorrhization. *African Journal of Biotechnology*, 8(12), 2750-2756.

Koffi M C, Declerck S (2015) *In vitro* mycorrhization of banana (*Musa acuminata*) plantlets improves their growth during acclimatization. *In Vitro Cellular & Developmental Biology- Plant,* 51(3), 265-273.

Lalaymia I, Declerck S (2020) The Mycorrhizal Donor Plant (MDP) *in vitro* culture system for the efficient colonization of whole plants. 10th International Molecular Biology and Biotechnology Congress (04-08 Oct. 2021) Proceedings Book 134. *Arbuscular Mycorrhizal Fungi: Methods and Protocols*, 19-31.

Lotfi M, Fernandez K, Vermeir P, Mars M, Werbrouck S (2019b) *In vitro* mycorrhization of pear (*Pyrus communis*). *Mycorrhiza*, 29(6), 607-614.

Lotfi M, Mars M, Werbrouck S (2019a) Optimizing pear micropropagation and rooting with light emitting diodes and trans-cinnamic acid. *Plant Growth Regulation, 88*(2), 173-180.

Louche-Tessandier D, Samson G, Hernandez-Sebastia C, Chagvardieff, P, Desjardins Y (1999) Importance of light and CO_2 on the effects of endomycorrhizal colonization on growth and photosynthesis of potato plantlets (*Solanum tuberosum*) in an *in vitro* tripartite system. *New Phytologist*, 142(3), 539-550.

Ouledali S, Ennajeh M, Zrig A, Gianinazzi S, Khemira H (2018) Estimating the contribution of arbuscular mycorrhizal fungi to drought tolerance of potted olive trees (*Olea europaea*). *Acta Physiologiae Plantarum*, 40(5), 81.

Rhouma A, Helali F, Chettaoui M, Hajjouji M, Hajlaoui MR (2014) First report of fire blight caused by *Erwinia amylovora* on pear in Tunisia. *Plant Disease*, 98(1), 158-158.

Shishikura M, Takemura Y, Sotome K, Maekawa N, Nakagiri A, Endo N (2021) Four mycelial strains of *Entoloma clypeatum* species complex form ectomycorrhiza-like roots with *Pyrus betulifolia* seedlings *in vitro*, and one develops fruiting bodies 2 months after inoculation. *Mycorrhiza*, 31(1), 31-42.

Sosa-Rodriguez T, de Boulois H D, Granet F, Gaurel S, Melgarejo L M, Carron M P, Declerck S (2013) *In vitro* mycorrhization of the rubber tree *Hevea brasiliensis* Müll Arg. *In Vitro Cellular & Developmental Biology-Plant*, 49(2), 207-215.

Varma A, Schüepp H (1995) Mycorrhization of the commercially important micropropagated plants. *Critical Reviews in Biotechnology*, 15(3- 4), 313-328.

Voets L, de la Providencia IE, Declerck S (2006) Glomeraceae and Gigasporaceae differ in their ability to form hyphal networks. *New Phytologist*, 172(2), 185-188.

Voets L, de la Providencia IE, Fernandez K, IJdo M, Cranenbrouck S, Declerck S (2009) Extraradical mycelium network of arbuscular mycorrhizal fungi allows fast colonization of seedlings under *in vitro* conditions. *Mycorrhiza,* 19(5), 347-356.

Voets L, Dupré de Boulois H, Renard L, Strullu D G, Declerck S (2005) Development of an autotrophic culture system for the *in vitro* mycorrhization of potato plantlets. *FEMS Microbiology Letters,* 248(1), 111-118.

Chapter 8

Nutrient Mineralization: A Function of Mycorrhizal Fungi Necessary for Plant Survival

Oladipo Abiodun[1,*]
Muse Saad[1]
Eko Omobola[2]
and Adewuyi Abiola[2]

[1] Co-Innovation Center for Sustainable Forestry in Southern China, College of Forestry, Nanjing Forestry University, Nanjing, China
[2] Department for Innovation in Biological, Agrofood and Forest Systems (DIBAF), University of Tuscia, Viterbo, Italy

Abstract

Mycorrhizal fungi are soil dwelling organisms that have a long history of association with plants. Through their hyphae, they are able to attract nutrient in distant soil for the benefit of plant growth. The symbiotic relationship between mycorrhizal fungi and plants facilitates the survival both plants and fungi. Nutrient availability and uptake are crucial for plants growth and development. However, available usable nutrients by plants are limited or depleted in different soil condition.

One sure way by which plants are able to get useable nutrient from the soil is associated with mycorrhizal fungi. Nutrients such as Phosphorus and Nitrogen in most cases are not available in usable form for plant uptake, but through the interaction of mycorrhizal fungi and other microbes, the mineralization of these elements have been made

* Corresponding Author's Email: oladipoabiodun@njfu.edu.cn.

In: New Research on Mycorrhizal Fungus
Editors: Qiang-Sheng Wu, Ying-Ning Zou, Yue-Jun He et al.
ISBN: 979-8-88697-637-3

possible. Therefore, we consider the role of mycorrhizal fungi in nutrient mineralization for the growth, development and survival of plants.

Keywords: soil, nutrient, mineralization, mycorrhizal fungi, plant growth

1. Introduction

The importance of plant generally cannot be over-emphasized considering the range of benefits they provide. One of such benefit is the ecosystem services and aesthetic architecture provided especially by ornamental plants. Improvement of urban and peri-urban environment cannot be separated from greenery because it contribute to favorable climate, carbon sequestration, environmental air purification, soil stabilization, noise reduction, supply of breathable oxygen and many more (Bolund and Hunhammar 1999; Saifuddin and Osman 2014). The ecosystem services provided by plants have necessitated several planting programs, contributing to the aesthetic beautification of the environment among other benefits (Joye et al., 2010). Scientific evidence further proves that greenery positively contribute to human well-being with significant enhancement of human psychological and cognitive abilities (Joye et al., 2010).

However, the survival of these unique plants is being challenged by several factors such as lack of proper maintenance practices (Shober, Denny, and Broschat 2010), soil/growth medium (Sari and Küçük 2019; Thakur, Sharma, and Sachin 2020), soil pH (Zhao et al., 2013), pathogenic attack (van Dijk, Ehrlén, and Tack 2021), and in particular, by change in climatic condition (De 2018) which is occurring at a global scale. One major environmental stress that generally affects the survival of plant is drought, and it is prevalent in almost every environment, mostly without any prior notification and it negatively affects quality, biomass and energy of plant (Seleiman et al., 2021); this is because fresh plant biomass is made up of about 80-95% of water (Brodersen et al., 2019). Drought is associated with low rainfall, high temperature and increased sunlight intensity leading to the disruption of plant's biochemical, morphological, molecular and physiological features (Oladipo et al., 2021; Seleiman et al., 2021).

The nutritional and water requirements of plants especially ornamentals, varies from species to species and it is associated with the health and visual quality of the plant. Availability of nutrient and other environmental factors contribute to the shape, color and height of ornamental plants. The application of fertilizers (organic or chemical) could enhance the overall health of these plants, however, lack of nutrients or improper application of fertilizers could result in low quality plants and financial loss due to high cost of production (Neto, Boldrin, and Mattson, 2015). The use of ornamental plants cuts across different environment from interior (homes and offices) to exterior (landscaping, gardening, and cut flowers for different events such as weddings, burials and other events) beautification. It has been reported that approximately one-third of the value of floriculture globally originates from cut flowers and the ornamental plants market globally is worth above 300 billion United State Dollars (Azadi et al., 2016). In addition, some flowers are edible and have been reported to contribute to the nutritional requirement of human diet (Mlcek and Rop 2011). Plant nutrient are basically chemical elements classified as micronutrients (copper, boron, chlorine, manganese, nickel, zinc, molybdenum, and iron) and macronutrients (calcium, potassium, phosphorus, magnesium, nitrogen and sulphur), nevertheless, plant's metabolism and lifecycle requires other elements such as carbon, hydrogen and oxygen (Kumar, Kumar, and Mohapatra 2021; Neto et al., 2015).

Arbuscular mycorrhizal fungi (AMF) are soil-borne fungi that can boost plant nutrient uptake (Figure 1) and resilience to a variety of abiotic stressors (Sun et al., 2018). AMF help host plants grow better under difficult conditions by mediating a wide range of complex interactions between the plant and the fungus, resulting in increased photosynthetic rate and other gas exchange-related traits, as well as higher water absorption (Birhane et al., 2012). There are majorly two types of mycorrhizal fungi that establish symbioses with the roots of most terrestrial plants; these are arbuscular mycorrhizal (AM) fungi and ectomycorrhizal (EM) fungi (Tedersoo 2017). AM-associated plants, in particular, rely on inorganic soil nitrogen (N) resources and dominate rapidly N cycling ecosystems with increased soil N mineralization rates, whereas EM-associated plants thrive in reduced N cycling ecosystems with poor soils N mineralization rates, owing to EM fungi's ability to break down and absorb organic N from soil (Bunn et al., 2019; Shah et al., 2016). Therefore, a holistic approach aimed at sustainable management of plants through nutrient mineralization by mycorrhizal fungi cannot be ignored. This chapter, therefore, focuses on the mineralization of nutrients especially from a biological view point, which is one of the functions of mycorrhizal fungi.

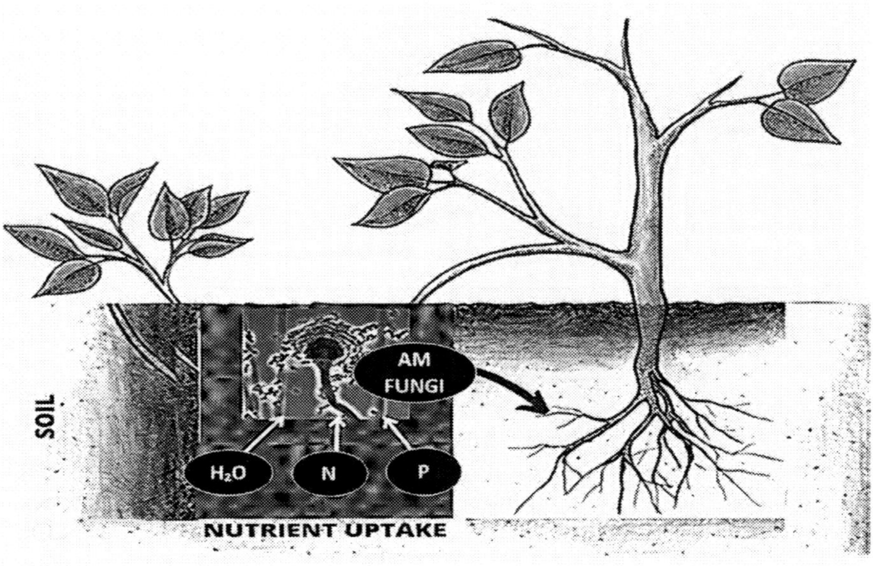

Figure 1. Arbuscular mycorrhizal fungi facilitates nutrient uptake in plant and contributes to overall plant growth and development.

2. Nutrient Mineralization

The process employed for the transformation (decomposition or oxidation) of chemicals present in organic matter or compounds into simpler forms that is readily available for plant's utilization can be described as mineralization (Grzyb, Wolna-Maruwka, and Niewiadomska 2020; Sharma and Garg 2018). One of the most significant natural resources is soil, which serves as a living reservoir for a wide range of flora and fauna and serves as a filtration system for the movement of water, minerals, and energy within an ecosystem. When it comes to ecosystem activities like the breakdown of organic matter, the recycling of nutrients, and the alteration of soil properties, soil flora and fauna are crucial players (Doran 2002; Schoonover and Crim 2015). The most numerous and diverse creatures on earth are soil microbial communities which play an essential role in decomposition of organic matter, nutrient distribution and formation of soil humic compounds; their biomass accounts for 5% of the soil's organic matter (Grzyb et al., 2020; Jamir et al., 2019).

Earthworms and other members of the soil microbiota, including fungi and bacteria, play a major role in the transformation of organic molecules in the soil (Doran 2002). According to Bassanino et al., (2011), sustainable agro-production and ecosystem balance depend on healthy and sustainable soil practices. High biodiversity, stress tolerance, and a balanced nutrient cycle are indicators of good soil. The mesophilic temperature encourages the heterotrophic and diverse microbial community, which improves mineralization through the metabolism of carbon molecules such soluble sugars, organic acids, and other similar chemicals (Gupta, Gupta, and Singh 2017). The process of mineralization is influenced by soil texture and structure, weather, the chemical make-up of organic matter, and human activities like fertilizer and ploughing (Vilkienė et al., 2016). The mineralization process is accelerated through the mixture of organic materials, mineral particles, and microbes, giving rise to new contact surfaces for the bacteria when mixed together (Polgári et al., 2019). In the work of Grzyb et al., (2020) it was stated that 20–25 percent of crop wastes go through humification, which results in the production of soil humus, while the remaining 75–80 percent are mineralized into nutrients that are available to plants; also the amount and activity of soil microbial communities is associated with the quantity, quality and distribution of organic matter.

Three reactions (aminization, ammonification, and nitrification) convert nitrogen during mineralization into inorganic forms (NH_4, NO_2, and NO_3) for the use of plants (Zhang, Dumroese, and Pinto 2019). The activity of the microorganisms and their needs for carbon and nitrogen, as well as the concentrations of dissolved organic nitrogen and ammonium, control these reactions (Sharma and Garg 2018; Sinha 2011). Through changes to the environment and interactions with microbes, earthworms also have a significant impact on nitrogen transformations during vermicomposting. Earthworms speed up nitrogen mineralization, which creates favorable nitrification conditions in organic wastes and causes ammonium-nitrogen to quickly transform into nitrates (Xia, Cui, and Li 2016).

Nitrogen (N) inputs into the soil have been shown to have a major impact on aboveground ecosystem productivity as well as belowground pools and fluxes of N such as microbial biomass N and mineralization (Changhui et al., 2014), therefore, survival of organisms in unfertilized soils is defined by significant nutrient competition among plant communities as well as between plant roots and microbes (Adler et al., 2018). Plant roots engage with rhizosphere bacteria to speed up the mineralization of soil organic matter (SOM) for nutrient uptake and changes in SOM mineralization caused by roots are heavily influenced by root-derived carbon (root-C) intake and soil nutrient condition. When C-to-N mineralization ratio increases with N availability, it is an indication that microorganisms acting on nitrogen rich soil could facilitate nitrogen mineralization. Plants, therefore in their effort to compete effectively for nutrient acquisition, increase specific root lengths (Sun et al., 2021).

In an experiment conducted by Xiao et al., (2022) to investigate the effect of forest understory vegetation of nitrogen mineralization and bacterial community, it was reported that there exist variation in N availability and mineralization rates, and microbial biomass was significant in 0-10 cm soil layer compared to 10-20 cm soil layer. Further, the correlation between microbial biomass and soil nitrogen mineralization was stronger than that of soil nitrogen content. Therefore, understanding the mechanism of soil nitrogen dynamics, factors ranging from soil, climate, anthropogenic and microbes are essential.

3. Nutrient Mineralization by Mycorrhiza Fungi for Plant Use

The arbuscular mycorrhizal (AM) fungi coexist with more than eighty percent of terrestrial plant species, as stated by Smith and Smith (2011). These fungi are known to thrive outside the rhizosphere and form an extensive network of fine hyphae that penetrates soil pores that are inaccessible to roots, thereby gaining access into nutrient patches. AMF and plants live in symbiotic relationship where sucrose is released into peri-arbuscular space and hydrolyzed to produce fructose and glucose. Although it is possible that both hexoses might be transferred within the arbuscules, however, this is not certain (Parniske 2008; Lin Zhang, Feng, and Declerck 2018). AMF's intra-radical hyphae can directly absorb glucose and fructose from the root exudates, during the intra-radical hyphal stage, part of the hexoses is transformed into glycogen before being transferred to the extra-radical hyphal stage where trehalose is formed (Bago, Pfeffer, and Shachar-Hill 2000).

Phosphorus (P) availability varies greatly amongst soil types, influencing P nutrition in forest plants. Chemical P binding to soil moieties influences P partitioning between soil particles and solution, hence regulating soluble P concentrations on which plants, their associated mycorrhizal symbionts, and microorganisms depend (Clausing and Polle 2020). In organic materials, AM fungal hyphae proliferate in response to soil nutrient heterogeneity (Hodge et al., 2009). It is well known that AM fungal hyphae can obtain P (Figure 1) from organic P-rich patches and improve plant P nutrition (Tarafdar and Marschner 1994). Also, in microcosms, PSB has been shown to stimulate the growth of AM hyphae and increase organic P mineralization under a well-managed conditions. Organic P mineralization in the field has been improved by AM fungi with bacteria on the hyphal surface and alkaline phosphatases (alp) genes on their surface (Zhang et al., 2018). A broad root absorption area permits the roots to intensify the interface contact with soil hyphae, increasing the likelihood of attracting symbiotic AM fungus. Higher levels of AM colonization can be detected in heterogeneous soil environments compared to homogeneous soil environments (Liang et al., 2022; Martínez-García and Pugnaire 2011) which may improve nutrient uptake for plant root growth. Importantly, AM hyphae can penetrate between rocks to access regions where plant roots cannot (Nadian et al., 1998). This property is a very efficient technique of producing absorptive surfaces for plant development in karst locations with a high rock-soil ratio (Liang et al., 2022).

In the work of Clausing and Polle (2020), in which P uptake by mycorrhizal and non-mycorrhizal root tips were examined in P-rich and P-poor forest considering soil cores at 20 cm depth. It was observed that in P-poor forest, root P uptake was 60% in organic layer and 40% in mineral layer while in P-rich forest, 25% and 75% were observed for organic and mineral layer, however, lower enrichment was experienced for non-mycorrhiza root tips. In the examination of P uptake in *Capsicum annuum* by AM fungi using 0, 10, and 200 mg kg^{-1}. The result revealed that P uptake by shoot significantly increased by function of AM fungi at 0 mg kg^{-1} and 10 mg kg^{-1}. With about 50% of inoculated plant roots infected with AM fungi, external hyphae recorded 20 m g^{-1} soil at 10 and 200 mg kg^{-1}. It can be argued that the high efficacy of hyphal surfaces in absorbing P from soil may be sufficient to explain the favorable effect of AM on P uptake from soil in most circumstances (Sharif and Claassen 2011).

In another research (Navarro and Morte 2019), where seedling of *Citrus macrophylla* employed to examine the effect P nutrition having been inoculated with a mixture of two different mycorrhiza fungi (*Rhizophagus irregularis* and *Funneliformis mosseae*) and treated with 0, 0.1, 1, and 5 mM of phosphorus for 165 days. Observation revealed that AM played a significant role in P uptake and plant growth which was a function of P fertilization, while P nutrition in AM plants was significantly improved with highest growth recorded at 1 mM of phosphorus. However, AM plants receiving highest P supply did not necessarily result in enhanced mineral nutrition when compared to uninoculated plants (Navarro and Morte 2019). The effect of *Funneliformis mosseae* on soil organic P mineralization and P absorption in *Camellia oleifera* after 4 and 8 months of planting was studied. As reported, *F. mosseae* inoculation boosted soil acid phosphatase, phytase, and alkaline phosphatase activities while decreasing soil organic P content. Regardless of organic P treatment, mycorrhizal plants had higher root activity, shoot and root P levels, and root acid phosphatase and phytase activities than non-mycorrhizal plants. Finally, AM fungus improved soil organic P mineralization and influenced root phosphatase activity (Wu et al., 2021). Although, AM fungi, play an important role in mineralization and uptake of nutrient, it should be well noted that interaction with other microbes also influences the whole process (Jiang et al., 2021; Nacoon et al., 2020).

It should also be noted that N-limited plants are more associated with new soil in high-latitude boreal-arctic ecosystems while P-limited soil tends to be associated with weathered old soil in low-latitude tropical forests. Assuming that AM plants are less nutrient-conservative compared to EM plants, and the major element reducing the growth of plant differs from tropic to poles, it is possible that, at higher latitudes EM plant will conserve more N and at low latitudes will conserve more P. however, it is also possible for EM plants to conserve more N and P than AM plants across all latitudes if the mechanism of N and P are connected to the general mechanism of plant nutrition (Colin et al., 2019).

Conclusion and Future Prospects

The interconnectedness between plants, soil, mycorrhizal fungi and mineralization is a complex one that should be unveiled for optimum growth and development of plants under different ecosystem and climatic condition. Although, mineralization involves other microorganisms, the role of mycorrhizal fungi cannot be ignored because, apart from facilitating mineralization process, accessibility of plants to mineralized nutrient is a key function of these unique soil-dwelling fungi. It is therefore necessary to understand the dynamics of all components in this complex relationship because the effectiveness of mycorrhizal fungi not only in the process of mineralization but also in nutrient supply for plant's use is possibly being influenced by other microorganism, soil condition, climatic factors as well as the plant in question. Understanding the uniqueness of such complexities is a possible approach for result-oriented application of mycorrhizal fungi.

Acknowledgments

All authors hereby acknowledge the support of Qiang-Sheng Wu, who provided informative guide and correction towards the success of this chapter.

References

Adler, P. B., Danielle Smull, Karen H. Beard, Ryan T. Choi, Tucker Furniss, Andrew Kulmatiski, Joan M. Meiners, Andrew T. Tredennick, and Kari E. Veblen. 2018. "Competition and Coexistence in Plant Communities: Intraspecific Competition Is Stronger than Interspecific Competition." *Ecology Letters* 21(9):1319–29. doi: https://doi.org/10.1111/ele.13098.

Azadi, P., Hedayat Bagheri, Ayoub Molaahmad Nalousi, Farzad Nazari, and Stephen F. Chandler. 2016. "Current Status and Biotechnological Advances in Genetic Engineering of Ornamental Plants." *Biotechnology Advances* 34(6):1073–90. doi: https://doi.org/10.1016/j.biotechadv.2016.06.006.

Bago, B., P. E. Pfeffer, and Y. Shachar-Hill. 2000. "Carbon Metabolism and Transport in Arbuscular Mycorrhizas." *Plant Physiology* 124(3):949–58. doi: https://doi.org/10.1104/pp.124.3.949.

Bassanino, M., Dario Sacco, Laura Zavattaro, and Carlo Grignani. 2011. "Nutrient Balance as a Sustainability Indicator of Different Agro-Environments in Italy." *Ecological Indicators - ECOL INDIC* 11:715–23. doi: https://doi.org/10.1016/j.ecolind.2010.05.005.

Birhane, E., Frank J. Sterck, Masresha Fetene, Frans Bongers, and Thomas W. Kuyper. 2012. "Arbuscular Mycorrhizal Fungi Enhance Photosynthesis, Water Use Efficiency, and Growth of Frankincense Seedlings under Pulsed Water Availability Conditions." *Oecologia* 169(4):895–904. doi: https://doi.org/10.1007/s00442-012-2258-3.

Bolund, P., and Sven Hunhammar. 1999. "Ecosystem Services in Urban Areas." *Ecological Economics* 29(2):293–301. doi: https://doi.org/10.1016/S0921-8009(99)00013-0.

Brodersen, C. R., Adam B. Roddy, Jay W. Wason, and Andrew J. McElrone. 2019. "Functional Status of Xylem Through Time." *Annual Review of Plant Biology* 70(1):407–33. doi: https://doi.org/10.1146/annurev-arplant-050718-100455.

Bunn, R. A., Dylan T. Simpson, Lorinda S. Bullington, Ylva Lekberg, and David P. Janos. 2019. "Revisiting the 'Direct Mineral Cycling' Hypothesis: Arbuscular Mycorrhizal Fungi Colonize Leaf Litter, but Why?" *The ISME Journal* 13(8):1891–98. doi: https://doi.org/10.1038/s41396-019-0403-2.

Changhui, W., Zhu Feng, Zhao Xiang, and Dong Kuanhu. 2014. "The Effects of N and P Additions on Microbial N Transformations and Biomass on Saline-Alkaline Grassland of Loess Plateau of Northern China." *Geoderma* 213:419–25. doi: https://doi.org/10.1016/j.geoderma.2013.08.003.

Clausing, S., and Andrea Polle. 2020. "Mycorrhizal Phosphorus Efficiencies and Microbial Competition Drive Root P Uptake." *Frontiers in Forests and Global Change* 3. doi: https://doi.org/10.3389/ffgc.2020.00054.

Colin, A., Bhatnagar Jennifer M., Dietze Michael C., Pearse William D., and Kivlin Stephanie N. 2019. "Global Imprint of Mycorrhizal Fungi on Whole-Plant Nutrient Economics." *Proceedings of the National Academy of Sciences* 116(46):23163–68. doi: https://doi.org/10.1073/pnas.1906655116.

De, L. 2018. "*Impact of Climate Change on Floriculture and Landscape Gardening.*" 10:6253–56.

Doran, J. W. 2002. "Soil Health and Global Sustainability: Translating Science into Practice." *Agriculture, Ecosystems & Environment* 88(2):119–27. doi: https://doi.org/10.1016/S0167-8809(01)00246-8.

Grzyb, A., Agnieszka Wolna-Maruwka, and Alicja Niewiadomska. 2020. "Environmental Factors Affecting the Mineralization of Crop Residues." *Agronomy* 10(12).

Gupta, A., R. Gupta, and R. L. Singh. 2017. "Microbes and Environment." *Principles and Applications of Environmental Biotechnology for a Sustainable Future* 43–84.

Hodge, A., Graziella Berta, Claude Doussan, Francisco Merchan, and Martin Crespi. 2009. "Plant Root Growth, Architecture and Function." *Plant and Soil* 321(1):153–87. doi: https://doi.org/10.1007/s11104-009-9929-9.

Jamir, E., Rajiv Das Kangabam, Kasmika Borah, Abhijit Tamuly, H. P. Deka Boruah, and Yumnam Silla. 2019. "*Role of Soil Microbiome and Enzyme Activities in Plant Growth Nutrition and Ecological Restoration of Soil Health BT - Microbes and Enzymes in Soil Health and Bioremediation.*" Pp. 99–132 in, edited by A. Kumar and S. Sharma. Singapore: Springer Singapore.

Jiang, F., Lin Zhang, Jiachao Zhou, Timothy S. George, and Gu Feng. 2021. "Arbuscular Mycorrhizal Fungi Enhance Mineralisation of Organic Phosphorus by Carrying Bacteria along Their Extraradical Hyphae." *New Phytologist* 230(1):304–15. doi: https://doi.org/10.1111/nph.17081.

Joye, Y., Kim Willems, Malaika Brengman, and Kathleen Wolf. 2010. "The Effects of Urban Retail Greenery on Consumer Experience: Reviewing the Evidence from a Restorative Perspective." *Urban Forestry & Urban Greening* 9(1):57–64. doi: https://doi.org/10.1016/j.ufug.2009.10.001.

Kumar, S., Santosh Kumar, and Trilochan Mohapatra. 2021. "Interaction Between Macro- and Micro-Nutrients in Plants." *Frontiers in Plant Science* 12. doi: https://doi.org/10.3389/fpls.2021.665583.

Liang, Y., Fujing Pan, Zhongcheng Jiang, Qiang Li, Junbing Pu, and Kunping Liu. 2022. "Accumulation in Nutrient Acquisition Strategies of Arbuscular Mycorrhizal Fungi and Plant Roots in Poor and Heterogeneous Soils of Karst Shrub Ecosystems." *BMC Plant Biology* 22(1):188. doi: https://doi.org/10.1186/s12870-022-03514-y.

Martínez-García, L. B., and Francisco I. Pugnaire. 2011. "Arbuscular Mycorrhizal Fungi Host Preference and Site Effects in Two Plant Species in a Semiarid Environment." *Applied Soil Ecology* 48(3):313–17. doi: https://doi.org/10.1016/j.apsoil.2011.04.003.

Mlcek, J., and Otakar Rop. 2011. "Fresh Edible Flowers of Ornamental Plants – A New Source of Nutraceutical Foods." *Trends in Food Science & Technology* 22(10):561–69. doi: https://doi.org/10.1016/j.tifs.2011.04.006.

Nacoon, S., Sanun Jogloy, Nuntavun Riddech, Wiyada Mongkolthanaruk, Thomas W. Kuyper, and Sophon Boonlue. 2020. "Interaction between Phosphate Solubilizing Bacteria and Arbuscular Mycorrhizal Fungi on Growth Promotion and Tuber Inulin Content of Helianthus Tuberosus L." *Scientific Reports* 10(1):4916. doi: https://doi.org/10.1038/s41598-020-61846-x.

Nadian, H., S. E. Smith, A. M. Alston, R. S. Murray, and B. D. Siebert. 1998. "Effects of Soil Compaction on Phosphorus Uptake and Growth of Trifolium Subterraneum Colonized by Four Species of Vesicular–Arbuscular Mycorrhizal Fungi." *New Phytologist* 140(1):155–65. doi: https://doi.org/10.1046/j.1469-8137.1998.00219.x.

Navarro, J. M., and Asunción Morte. 2019. "Mycorrhizal Effectiveness in Citrus Macrophylla at Low Phosphorus Fertilization." *Journal of Plant Physiology* 232:301–10. doi: https://doi.org/10.1016/j.jplph.2018.11.027.

Neto, A., Karina Boldrin, and Neil Mattson. 2015. "Nutrition and Quality in Ornamental Plants." *Revista Brasileira de Horticultura Ornamental* 21:139. doi: https://doi.org/10.14295/aohl.v21i2.809.

Oladipo, A., Ademola Adebayo, Olakunle Ogunyemi, Adegboyega Aduradola, and Fu Fang-Fang. 2021. "*Effects of Mycorrhizae Inoculation and Water Regime on the Growth of Cassia Fistula L. Seedlings*." 27:483–95.

Parniske, M.. 2008. "Arbuscular Mycorrhiza: The Mother of Plant Root Endosymbioses." *Nature Reviews Microbiology* 6(10):763–75. doi: https://doi.org/10.1038/nrmicro1987.

Polgári, M., Ildikó Gyollai, Krisztián Fintor, Henrietta Horváth, Elemér Pál-Molnár, and João Carlos Biondi. 2019. "Microbially Mediated Ore-Forming Processes and Cell Mineralization." *Frontiers in Microbiology* 10:2731. doi: https://doi.org/10.3389/fmicb.2019.02731.

Saifuddin, M., and Normaniza Osman. 2014. "Evaluation of Hydro-Mechanical Properties and Root Architecture of Plants for Soil Reinforcement." *Current Science* 107(5):845–52.

Sari, D., and Mehmet küçük. 2019. "Analysis of the Effects of Certain Exotic Ornamental Plants on Soil Properties: the Karadeniz Technical University Campus Case." *International Journal of Ecosystems and Ecology Science (IJEES)* 9:447–60. doi: https://doi.org/10.31407/ijees9306.

Schoonover, J. E., and Jackie F. Crim. 2015. "An Introduction to Soil Concepts and the Role of Soils in Watershed Management." *Journal of Contemporary Water Research & Education* 154(1):21–47. doi: https://doi.org/10.1111/j.1936-704X.2015.03186.x.

Seleiman, M. F., Nasser Al-Suhaibani, Nawab Ali, Mohammad Akmal, Majed Alotaibi, Yahya Refay, Turgay Dindaroglu, Hafiz Haleem Abdul-Wajid, and Martin Leonardo Battaglia. 2021. "Drought Stress Impacts on Plants and Different Approaches to Alleviate Its Adverse Effects." *Plants (Basel, Switzerland)* 10(2):259. doi: https://doi.org/10.3390/plants10020259.

Shah, F., César Nicolás, Johan Bentzer, Magnus Ellström, Mark Smits, Francois Rineau, Björn Canbäck, Dimitrios Floudas, Robert Carleer, Gerald Lackner, Jana Braesel, Dirk Hoffmeister, Bernard Henrissat, Dag Ahrén, Tomas Johansson, David S. Hibbett, Francis Martin, Per Persson, and Anders Tunlid. 2016. "Ectomycorrhizal Fungi Decompose Soil Organic Matter Using Oxidative Mechanisms Adapted from

Saprotrophic Ancestors." *New Phytologist* 209(4):1705–19. doi: https://doi.org/10.1111/nph.13722.

Sharif, M. and N. Claassen. 2011. "Action Mechanisms of Arbuscular Mycorrhizal Fungi in Phosphorus Uptake by *Capsicum annuum* L." *Pedosphere* 21(4):502–11. doi: https://doi.org/10.1016/S1002-0160(11)60152-5.

Sharma, K., and Vinod Kumar Garg. 2018. "Chapter 17 - Solid-State Fermentation for Vermicomposting: A Step Toward Sustainable and Healthy Soil." *Current Developments of Biotechnology and Bioengineering*, pp. 373–413, edited by A. Pandey, C. Larroche, and C. R. B. T. C. D. in B. and B. Soccol. Elsevier.

Shober, A. L., Geoffrey C. Denny, and Timothy K. Broschat. 2010. "Management of Fertilizers and Water for Ornamental Plants in Urban Landscapes: Current Practices and Impacts on Water Resources in Florida." *HortTechnology Hortte* 20(1):94–106. doi: https://doi.org/10.21273/HORTTECH.20.1.94.

Sinha, R. 2011. "Organic Farming by Vermiculture: Producing Safe, Nutritive and Protective Foods by Earthworms (Charles Darwin's Friends of Farmers)." *American Journal of Experimental Agriculture* 1:363–99. doi: https://doi.org/10.9734/AJEA/2011/519.

Smith, F., and Sally Smith. 2011. "What Is the Significance of the Arbuscular Mycorrhizal Colonisation of Many Economically Important Crop Plants? Plant Soil." *Plant and Soil* 348:63–79. doi: https://doi.org/10.1007/s11104-011-0865-0.

Sun, Y., Huadong Zang, Thomas Splettstößer, Amit Kumar, Xingliang Xu, Yakov Kuzyakov, and Johanna Pausch. 2021. "Plant Intraspecific Competition and Growth Stage Alter Carbon and Nitrogen Mineralization in the Rhizosphere." *Plant, Cell & Environment* 44(4):1231–42. doi: https://doi.org/10.1111/pce.13945.

Sun, Z., Jiabin Song, Xi'an Xin, Xianan Xie, and Bin Zhao. 2018. "Arbuscular Mycorrhizal Fungal 14-3-3 Proteins Are Involved in Arbuscule Formation and Responses to Abiotic Stresses During AM Symbiosis." *Frontiers in Microbiology* 9. doi: https://doi.org/10.3389/fmicb.2018.00091.

Tarafdar, J. C., and Horst Marschner. 1994. "Efficiency of VAM Hyphae in Utilisation of Organic Phosphorus by Wheat Plants." *Soil Science and Plant Nutrition* 40(4):593–600. doi: https://doi.org/10.1080/00380768.1994.10414298.

Tedersoo, L. 2017. "*Global Biogeography and Invasions of Ectomycorrhizal Plants: Past, Present and Future BT - Biogeography of Mycorrhizal Symbiosis*." Pp. 469–531 in, edited by L. Tedersoo. Cham: Springer International Publishing.

Thakur, N., Priyanshu Sharma, and T. M. Sachin. 2020. "Use of Alternative Growing Media in Ornamental Plants." *International Journal of Chemical Research* 8:2349–8528. doi: https://doi.org/10.22271/chemi.2020.v8.i6c.11079.

van Dijk, L. J. A., Johan Ehrlén, and Ayco J. M. Tack. 2021. "Direct and Insect-Mediated Effects of Pathogens on Plant Growth and Fitness." *Journal of Ecology* 109(7):2769–79. doi: https://doi.org/10.1111/1365-2745.13689.

Vilkienė, M., Dalia Ambrazaitienė, Danutė Karčauskienė, and Zenonas Dabkevičius. 2016. "Assessment of Soil Organic Matter Mineralization under Various Management Practices." *Acta Agriculturae Scandinavica, Section B - Soil & Plant Science* 66(8):641–46. doi: https://doi.org/10.1080/09064710.2016.1162845.

Wu, F., Zhengyun Li, Yulan Lin, and Linping Zhang. 2021. "Effects of *Funneliformis mosseae* on the Utilization of Organic Phosphorus in *Camellia oleifera* Abel." *Canadian Journal of Microbiology* 67(5):349–57. doi: https://doi.org/10.1139/cjm-2020-0227.

Xia, H., G. Cui, and F. Li. 2016. "Effects of Earthworms on Nitrification and Ammonia Oxidizers in Vermicomposting Systems for Recycling of Fruit and Vegetable Wastes." *Science of The Total Environment* 578. doi: https://doi.org/10.1016/j.scitotenv.2016.10.172.

Xiao, R., Xiuling Man, Beixing Duan, Tijiu Cai, Zhaoxin Ge, Xuefei Li, and Timo Vesala. 2022. "Changes in Soil Bacterial Communities and Nitrogen Mineralization with Understory Vegetation in Boreal Larch Forests." *Soil Biology and Biochemistry* 166:108572. doi: https://doi.org/10.1016/j.soilbio.2022.108572.

Zhang, L., Gu Feng, and Stéphane Declerck. 2018. "Signal beyond Nutrient, Fructose, Exuded by an Arbuscular Mycorrhizal Fungus Triggers Phytate Mineralization by a Phosphate Solubilizing Bacterium." *The ISME Journal* 12(10):2339–51. doi: https://doi.org/10.1038/s41396-018-0171-4.

Zhang, L., N. Shi, Jiequn Fan, Fei Wang, Timothy S. George, and Gu Feng. 2018. "Arbuscular Mycorrhizal Fungi Stimulate Organic Phosphate Mobilization Associated with Changing Bacterial Community Structure under Field Conditions." *Environmental Microbiology* 20(7):2639–51. doi: https://doi.org/https://doi.org/10.1111/1462-2920.14289.

Zhang, P., R. Kasten Dumroese, and Jeremiah R. Pinto. 2019. "Organic or Inorganic Nitrogen and Rhizobia Inoculation Provide Synergistic Growth Response of a Leguminous Forb and Tree." *Frontiers in Plant Science* 10. doi: https://doi.org/10.3389/fpls.2019.01308.

Zhao, D., Zhaojun Hao, Jing Wang, and Jun Tao. 2013. "Effects of PH in Irrigation Water on Plant Growth and Flower Quality in Herbaceous Peony (*Paeonia Lactiflora* Pall.)." *Scientia Horticulturae* 154:45–53. doi: https://doi.org/10.1016/j.scienta.2013.02.023.

Chapter 9

The Role of Arbuscular Mycorrhizal Fungi on Soil Health Improvement: A Component for Sustainable Agricultural Production

Selvaraj Anandakumar*
Thangavel Kalaiselvi
Sellappan Ranjith
and Dhandapani Senthamilselvi
Department of Agricultural Microbiology,
Tamil Nadu Agricultural University, Coimbatore, Tamil Nadu, India

Abstract

In the current scenario of agriculture, soil health or quality status is the major concern along with achieving food grain production to meet the world demand. Intensive use of inorganic chemicals and fertilizers heavily affects soil health, which leads to a significant reduction in agricultural production as a consequence of negative effects on macro and microfauna in the soil system. However, the world is searching for an eco-friendly technique to resolve and maintain soil health without harming the environment.

In agriculture, the use of plant growth promoting microorganisms (PGPMs), including arbuscular mycorrhizal (AM) fungi as an alternative and eco-friendly method to improve soil health and fertility. AM fungi are a unique soil fungus that are completely dependent on their host plants and have symbiotic association with most of the terrestrial plant species. This fungus improves soil health and fertility by regulating

* Corresponding Author's Email: anandasaval@gmail.com.

In: New Research on Mycorrhizal Fungus
Editors: Qiang-Sheng Wu, Ying-Ning Zou, Yue-Jun He et al.
ISBN: 979-8-88697-637-3

physiochemical and biological properties of soil with several mechanisms such as promoting soil aggregation, soil carbon and nutrients availability, water retention, microbial activities, nitrogen, carbon, and phosphorus cycling, thereby preventing soil erosion, carbon, and nitrogen loss as well as heavy metal remediation. Besides, AM fungi also offer sustainable benefits to associated plants such as nutrient availability, exclusively PO_4^{2-} and NO_3^-, improving defense mechanisms against phytopathogens and herbivorous insect pests as well as abiotic stresses drought, high temperature, heavy metal toxicity, and so on.

However, the effects of AMF on soil health and fertility are governed by the richness of AMF fungal species, tillage, and soil condition. This chapter helps to understand the significant role of AMF in enhancing soil health, and how it functions as a part of an integrated system for sustainable agriculture.

Keywords: arbuscular mycorrhizal fungi, soil health, nutrients, physio-chemical properties, biological properties, sustainable agriculture

1. Introduction

Agricultural production primly depends on soil health or quality, which provides essential services to sustain crop yield. Soil quality is defined as "the capacity of a soil to function within ecosystem boundaries to sustain biological productivity, preserve environmental quality and promote plant and animal health" (Doran and Parkin, 1994). It depends on the physio-chemical and biological properties of the soil (Schloter et al., 2018) that provide plants essential nutrients and water as well as contribute soil ecosystem services (Bedini et al., 2009). Moreover, healthy soil also supports a diverse and active community of biota, which contribute to element cycling, nutrient retention, and soil structure (Schloter et al., 2018). However, soil quality is continuously decreasing due to unsustainable land management practices like excessive use of inorganic fertilizers and chemicals, overgrazing, bush burning, continuous crop cultivations or intensive production, lower crop cover, and tillage practices as well as rapid civilization and industrialization (Dewitte et al., 2013).

Inoculation with arbuscular mycorrhizal fungi (AMFs) has been established as an eco-friendly approach of improving soil fertility (Powell and Rillig, 2018). The word "mycorrhiza" was invented by Frank (1885) to define and establish a symbiotic relationship with the roots of most terrestrial plants and fungi. The plethora of research noticed an increase in the growth and production of various field crops by root colonization of AM fungi *via* increasing the availability of P and other macro- and micro-nutrients including N, K, Mg, Cu, and Zn, especially in soils where they are present in less soluble forms (Smith and Read, 2008; Smith and Smith, 2012; Willis et al., 2013). Additionally, AM fungi also assist plants in obtaining water from the soil and enhance the soil structure. Mycorrhizae not only improve the availability of nutrients and water to plants, but also protects the plants from various abiotic stresses (Miransari, 2010).

Moreover, AM fungi improve the physical, chemical, and biological properties of the soil through carbon sequestration, nutrient retention, microbial activity, and soil structure. AM fungi have a positive impact on the soil by producing organic acids and glomalin, which prevent soil erosion, chelate heavy metals, improve carbon sequestration, and stabilize soil macro-aggregation. AMF are significant components of the soil microfauna and promote soil fertility by interacting with other rhizospheric microorganisms (Albrechtova et al., 2012). AMF also recruits bacteria that produce alkaline phosphatase, a mineralization soil enzyme associated with the available organic phosphorus. Furthermore, AMFs influence the composition, diversity, and activity of microbial communities in the soil through mechanisms of antagonism or cooperation. All of these AMF activities contribute to improving soil fertility.

Other benefits of AM fungi such as (1) facilitating plant access to moisture in dry soil conditions, (2) increasing soil carbon retention by preventing microbial degradation through enhanced soil particle aggregation and (3) creating a soil and rhizosphere environment that is more resilient to the development of plant disease may be co-benefits of more effective supply of nutrients to plants, but they can also exist independently in circumstances where the AM fungi have no particular role in nutrient-use efficiency.

However, the role of AM fungi is in turn influenced by the availability of soil phosphorus, the diversity and abundance of the AM fungi, AM fungal and host species, the size of the plant and its stage of development, the levels of soil carbon and nitrogen, application of mineral fertilizer, tillage and soil amendments (Ciccolini et al., 2015; Xu et al., 2017). Moreover, the impact of mycorrhizal colonization on soil health depends on the presence of soil bourn phytopathogen and other soil organisms (Endlweber and Scheu, 2007), soil salinity or acidity (Evelyn et al., 2009), heat and water limitation (Manoharan et al., 2010).

2. AM Fungi and Soil Biological Properties

Microorganisms are one of the most important soil components. These microorganisms interact with one another and the soil environment to contribute a variety of biochemical reactions and essential ecological processes including the solubilization of organic matter and the biogeochemical cycles of the elements, which have been proved to be essential ecosystem services for plant growth (Gessner et al., 2010; Nielsen et al., 2011). AM fungi are key components of soil microbiota and interact with other microorganisms in the rhizosphere by producing glomalin and the stimulation of root exudation into soils, which serves as a source of carbon to boost the soil microbiota (Bai et al., 2009). AM fungi influence the composition of the microflora in the rhizosphere by directly affecting the quantity and quality of root exudates (Baltrus, 2017). These changes affect both quantitatively and qualitatively the microbial populations in the rhizosphere of the mycorrhizal plant also termed the *mycorrhizosphere* (Gryndler, 2000).

The biological activities of AMF lead to the appearance of a positive, neutral, or negative relationship between AMFs and other soil microorganisms (Alimi et al., 2021). It is well recognized that beneficial microorganisms play fundamental roles in soil-plant systems. Particularly, plant growth promoting rhizobacteria (PGPR) exhibit a specific ability for root colonization and to improve plant growth, development and health (Gryndler, 2000; Barea et al., 2002). Numerous crucial ecosystem activities are carried out by the PGPR, including those involved in the biological control of plant diseases, nutrient cycling, seedling establishment, and soil quality (Barea, 2000). The positive interactions involve nutrient acquisition, biological control of root pathogens, improvement of plant tolerance to abiotic stresses, and soil fertility. AMFs interact with a wide range of microorganisms in the soil to enhance the soil fertility. It has been shown that the secretions of AMF influence the composition and activity of microbial communities in the rhizosphere (Veresoglou and Rillig, 2012).

Mycorrhizal fungi also interact with both symbiotic and free-living nitrogen (N_2)-fixing bacteria (Barea, 1997). The inoculation of mycorrhizal fungi has been shown to improve nodulation and N_2 fixation. A great deal of work has been carried out on the tripartite symbiosis legume (general term)-mycorrhiza-*Rhizobium* (general term) (Barea, 2000). Interactions between mycorrhizal fungi and *Azospirillum* have been reviewed by Volpin & Kapulnik (1994) and it has been demonstrated that *Azospirillum* could enhance mycorrhizal formation and response while mycorrhizal fungi may improve

Azospirillum establishment in the rhizosphere. The well-known activities of nitrogen-fixing bacteria and phosphate-solubilizing microorganisms improving the bioavailability of the major plant nutrients N and P are very much enhanced in the rhizosphere of mycorrhizal plants where synergistic interactions of such microorganisms with mycorrhizal fungi have been demonstrated. This is confirmed by Wang et al. (2011) and Xu et al. (2018) who demonstrated a synergistic relationship between AMF and the bacterial and fungal communities depending on N and P status in the soil. However, these interactions depend on several factors, such as the amounts of phosphorus and nitrogen available (Larimer et al., 2014).

3. AM Fungi and Soil Physical Properties

3.1. Soil Aggregation

Soil structure is commonly reflected as a measurement of soil aggregate strength and the main feature that controls physio-chemical and biological characteristics heading soil dynamics (Bronick and Lal, 2005). Soil aggregates are comprised of soil minerals (clay particles, fine sand and silt), plant roots, fungal hyphae, bacteria, free amorphous organic matter and organic matter intensely linked to clay coatings (Six et al., 2001). Soil aggregates are significant for retaining soil porosity, water filtration, plant and microbial growth, and improving strength against wind and water erosion (Six et al., 2001; Rillig et al., 2007).

AM fungi have a beneficial effect on soil structure. The AMF mycelia are present in massive quantities in soils (de Novais et al., 2019). These mycelia or hyphae have the property to create stable soil aggregations. Research outcomes showed that AM fungi help in the stabilization of soil aggregates by both AM hyphae and enhanced root growth due to AM fungi (Bearden and Petersen, 2000). Miller and Jastrow (1992) recommended effect of AM fungi on aggregation can be described as three distinct but simultaneous processes:

(1) Development of hyphae into soil matrix creating the thin structure that physically entangles primary soil particles,
(2) Roots and hyphae creating circumstances that enabling microaggregate formation in soil, and

(3) Roots and hyphae entangle resulting into the binding of microaggregates and small macroaggregates into bigger aggregates.

Tisdall et al. (1997) found that the efficiency of the fungal hyphae is determined by the hyphal length, surface area and their secretion of polysaccharides. The hyphal network of both mycorrhiza as well as saprophytic fungi, accompanied by fibrous roots, aggregates soil particles and microaggregates into bigger, aggregated entities (Miller and Jastrow, 1990). In addition, the mycelia network constantly renews itself and the dead mycelia also preserve soil structure until decomposition (Gianinazzi et al., 2010). These dead mycelia contribute to the stocks of organic matter and physical binders involved in soil aggregation (Hamel and Plenchette, 2017). All these mechanisms reduce the risks of soil compaction and promote soil fertility (Norton et al., 2020).

The important role of the extraradical mycelium in the formation of water-stable soil aggregates is well documented (Miller & Jastrow, 2000), and the involvement of glomalin, a glycoprotein produced by the external hyphae of mycorrhizal fungi, has been demonstrated (Wright & Upadhyaya, 1998). Glomalin has been suggested to contribute to the hydrophobicity of soil particles and also, because of its glue-like hydrophobic nature, to participate in the initiation of soil aggregates. Mycorrhizal fungi act as a long-term soil binding agent through the production of a glycoprotein (glomalin) by the extraradical mycelia (Singh et al., 2020). This glomalin is a hydrophobic, thermo-tolerant, or heat-tolerant resistant to the hot temperature of the soil. The hydrophobic character of the glomalin confers resistance of soil aggregations to water, and the production of this substance reaches its maximum in senescent mycelia.

The glycoprotein is slowly biodegradable by bacteria and fungi in the soil. The main function of glomalin is to stabilize soil aggregations (Hu et al., 2019; Mubekaphi, 2019), act as a glue that binds together the soil micro-aggregations (diameter <250μm) to form stable macro-aggregations (Lehmann et al., 2020). These soil macro-aggregations ensure better water infiltration, reduce surface runoff, control soil erosion, reduce nutrients and organic matter losses, increase gas exchange better retention of water and minerals, especially potassium, therefore, improve crop productivity (Demenois et al., 2018; Parihar et al., 2020). The improvement of soil aggregation contributes to the maintenance of good water infiltration rates, good tilth and adequate aeration for plant growth, thus improving soil quality (Wright & Upadhyaya, 1998). Moreover, microbial partners and plant roots of mycorrhizal association synthesize mucigels and polysaccharides that acts as gums and glues which bind and form steady aggregates (Oades, 1984).

Finally, it can be said that AMFs improve soil structure through their chemical and biophysical mechanisms, such as enmeshment and alignment. However, the effect of AM fungi on soil structure is influenced by tillage operations, which physically disturb soil aggregates and AM hyphal complexes. This act depreciates soil structure, and declines fertility and nutrient cycling capacity (Nichols and Wright, 2004).

4. AM Fungi and Soil Chemical Properties

4.1. Soil Organic Carbon and AM Fungi

Soil organic matter are having a significant role in soil fertility and productivity by exerting several positive effects on soil properties. Therefore, most agricultural soils show considerably lower SOC contents than soils with a natural and continuous vegetation cover (Sanderman et al., 2017). To overcome, key to adopt land use types and agricultural practices, such as use of cover crops, tillage, and application of organic and mineral fertilizers, modulate and maintain the amount of SOC and microbial abundance and activity, and increase the availability of nitrogen (N), potassium (K), phosphorus (P) and sulphur (S) in soil to crops over the long-term (Hallama et al., 2019; Oldfield et al., 2019). Soil properties such as soil organic matter, alkali-hydrolyzable N, available P, and pH were considerably positively linked with either total AM occupation or richness of spore and hyphae.

AM fungi play an essential role in the global C cycle. AMF hyphae are involved in C translocation into the soil and provide a key link in the terrestrial C cycle (Finlay, 2008). Indeed, AMF is an efficient agent to improve carbon sequestration in a mechanism of translocation of C away from the high respiratory activity around the root and into the soil aggregations (Zhu and Miller, 2003). Mycorrhizal association changes the quality of root secretion. The enhanced root secretion enhances soil microfloral population participating in organic matter decay and nutrient recycling. Dynamic nature of roots by diverse species of mycorrhizal fungi, and the wide range of functional responses of plants, present complex challenges for understanding all of the roles that mycorrhiza play in soil fertility (Abbott and Johnson, 2017). It has been demonstrated that mycorrhizal roots create a sink demand for carbon.

When the atmospheric CO_2 increases, the allocation of C from the plants to AMF also increases and stimulates the growth of AMF (Drigo et al., 2010). This C demand is provided by the host plant from the C fixed through photosynthesis (Parihar et al., 2020). In addition, AMF extramatrical hyphae represent 20–80% soil microbial biomass which consists of 15% of soil organic C (Leake et al., 2004). Moreover, AMF plays also a critical in the formation and maintenance of soil aggregations through the production of glomalin. This glomalin protects organic matter from microbial degradation, increases the hydrophobicity and stability of macro-aggregations, which control soil carbon loss and increase soil carbon stocks (Rillig et al., 2010).

4.2. AM Fungi and Soil Nutrient Availability

AM fungi are directly involved in elemental weathering, making soil nutrients available (Jurinak et al., 1986). It has also been demonstrated that mycorrhizal associations could play a significant role in the decomposition and mineralization of plant organic matter and mobilize nutrients, particularly nitrogen, for the benefit of the host plant (Lambers et al., 2008).

Interestingly, AM fungi play a key role in improving P availability in the soil by the process to transform P into bio-available forms via a range of chemical reactions and biological interactions (Zhu et al., 2018). It was believed that AMF hydrolyzes the organic P into inorganic phosphorus (Shen et al., 2011) through a mechanism linked to the production of enzymes named phosphatase (Tarafdar and Marschner, 1994), which cleave substrates present in the soil and make the phosphate accessible (Liu et al., 2013). However, recent studies revealed that AMF cannot release phosphatases into the soil (Zhang et al., 2016) but they recruit bacteria known as phosphate solubilizing bacteria (PSB) that produce phosphatase, which mineralizes organic P and provides a function that is absent from the AMF (Zhang et al., 2018; Etesami and Jeong, 2021).

PSB and AMF association is a beneficial feature that can mineralize insoluble phosphate in the soil and release soluble P that can easily be assimilated by plants (Wei et al., 2017; Mahanta et al., 2018). In that association, the role of PSB is to produce organic acids, such as gluconic acid, ketogluconic acids, siderophores, protons, and acid phosphatases that are involved in the mineralization of organic P in soil (Dobbelaere et al., 2003; Lucy et al., 2004), while AMF mycelia improve the absorption of soluble P in the plants (Taktek et al., 2017). The phosphatase releases P from organic P or inorganic orthophosphate by hydrolyzing phosphoric acid monoesters into P ion and a molecule with a free hydroxyl group (Othman and Panhwar, 2014). It was proven that the double inoculation of *Rhizophagus irregularis* and *Rahnella aquatilis* improves solubilization of inorganic P by the increased production of phosphatase released by the bacteria that is also stimulated by AMF exuded fructose (Zhang et al., 2018). AMF can also solubilize inorganic phosphate into soluble forms through the processes of acidification, chelation, exchange reactions, and production of organic acids, H^+, and metabolites (Behera et al., 2014). Moreover, the organic acid produced by AMF solubilizes insoluble mineral phosphate into a soluble form (Lapeyrie, 1988). AMF converts the insoluble P into soluble forms through the production of acids during their metabolic activities (Kalayu, 2019). But AM fungi do have both acid and alkaline phosphatases in both intraradical and extraradical hyphae (Aarle et al., 2002) and appear to have the ability to utilize P from phytates (Allen et al., 1981).

Additionally, the benefits included an increased N content, and a higher amount of organic matter and soil aggregation in the soil around the roots. It can be assumed that the increase in N content in the rhizosphere of the legume can be accounted for by an improvement in nodulation and N-fixation rates resulting from inoculation of nodulated plants with mycorrhizal fungi (Barea et al., 1992).

It is known that AMF allows better absorption of low mobile trace elements in soils, such as potassium (K), calcium (Ca), magnesium (Mg), copper (Cu), zinc (Zn), iron (Fe), manganese (Mn), and cobalt (Co) (Garcia et al., 2016; Hashem et al., 2018). For instance, according to Krishna and Bagyaraj (1984), the level of Zn, Fe, and Mn is twice in mycorrhizal peanut plants compared to non-mycorrhizal plants. It has also been revealed that mycorrhizal inoculation improved Zn and Cu nutrition in soybeans and clover (Schoeneberger et al., 1989). Mycorrhizal fungi are known to produce siderophores, high-affinity organic compounds that bind Fe^{3+} (Haselwandter, 2008) and then reduced to Fe^{2+} on the cell surface, which is transported into the fungus (Haselwandter et al., 2013). Moreover, AM fungi are known to improve Fe uptake and unavailable forms of soil nutrients. However, when some of these elements are present in high quantities and therefore possess a toxic character, mycorrhization can play a role in the protection of the plant, by strong retention of these elements (Liu et al., 2000).

5. Soil Stress and AM Fungi

5.1. Mitigation of Soil Heavy Metal Stress

Trace elements for instance Cu, Fe, Mn, Ni and Zn are critical for the natural growth and development of plants. These elements are needed in several enzymatic and oxidation-reduction reactions, and electron transfer and have a structural role in nucleic acid metabolism (Gohre and Paszkowski, 2006). Whereas some of the metals like Cd, Pb, Hg, and As are not vital for plant functions and even toxic even if present in small quantities within soils. High amounts of heavy metals obstruct vital enzymatic reactions by modification of protein structure or by substituting critical elements which ultimately results in deficiency symptoms. As a result, toxicity sign like chlorosis, growth delay and browning of roots can be detected (Garbisu and Alkorta, 2001). Measures like excavation and landfill, thermal treatment, acid leaching and electro reclamation are conventionally utilized for remediation of heavy metals but the practical applicability of these measures is limited due to their high cost, low effectiveness, great damage to soil structure and fertility (Saba et al., 2013). Microbes present in soil play a significant role in the mobilization and immobilization of metal cations, thus altering their accessibility to plants. Amongst soil microbes AMF are a very common inhabitant of soil present in nearly all types of soil environments and climates, playing a significant functional role in the soil-plant system (Barea et al., 1997) including disturbed soils (McGonigle and Miller, 1996).

Phytoremediation potential of plants can be improved by inoculation of mycorrhiza fungi as in improves heavy metal availability and plant's tolerance to heavy metals (Gaur and Adholeya, 2004), which collect, immobilize and detoxify heavy metals and safeguards plants from the high concentration of toxic heavy metals like copper, zinc, iron, manganese, cadmium, nickel, etc. The fungal hyphal network serves as a significant sink for heavy metals due to its large surface area of absorption. An alternate storage site for heavy metals is fungal vesicles. Moreover, AM fungi can improve plants' establishment and growth despite high concentrations of heavy metals in soil through better nutrition (Taylor and Harrier, 2001) and relieving abiotic stress (Auge, 2001). A plethora of reports confirmed that AMF might serve as an effective biotechnological strategy to help in the retrieval of heavy metal contaminated soils. Yang et al. (2017) studied the influence of arbuscular mycorrhizal fungi (AMF) on glomalin-related soil protein dissemination, aggregate stability and their relations with soil properties at various soil depths in lead and zinc polluted sites. Results showed that AMF played an important part in glomalin-related soil protein (GRSP), soil organic matter (SOM) and soil organic carbon (SOC) build-up and successively affected aggregate development and particle size dispersal in heavy metal contaminated soils.

5.2. Wasteland Reclamation

Presently one-fifth of the terrestrial ecosystems are under threat of desertification, which is visualized by disruption of natural plant communities but frequently goes together with or leads to the loss of vital physical, chemical and biological properties of soil (Skujins and Allen, 1986). Soil properties mainly govern soil worth and fertility which in turn supports the establishment and production of crops. Soil deprivation restricts its prospective for restoration of natural communities of plants. AMF showed great potential for retrieval of degraded soil and is usually utilized for reclamation of wastelands. The application of AM fungi can improve the growth and persistence of necessary re-vegetation species. Occupation by AMF can produce an advantageous physiological influence on the host plant in increasing the uptake of soil phosphorus. Nicolson (1967) proposed that plant growth in wastelands could be well improved by adding AMF. AM fungi improve a plant's capacity for the establishment and deal with stressed circumstances comprising nutrient shortage, drought, and soil disruption (Barea et al., 1997). Fungal mycelia spread from mycorrhizal roots by forming a three-dimensional net that connects plant roots and soil surroundings. It forms a resourceful structure for nutrient acquisition and searching in nutrient-deprived environments.

6. Factors Affecting Mycorrhizal Colonization and Soil Health

6.1. Impact of Crop Management on AMF and Soil

To maintain and improve nutrient availability and soil health and minimize pest and disease incidence, farmers use various management strategies, such as crop rotation and intercropping, to achieve sustainable agriculture production. These two practices influence the diversity, abundance, and functioning of AMF in the soil. For instance, crop rotation increases the number of AMF spore density and root colonization in maize, and it also increases wheat yield when the preceding crop is soybean or chickpea (Bakhshandeh et al., 2017). These positive effects are not the effects of leguminous nitrogen fixation but rather due to higher AMF activities in the soil (Bakhshandeh et al., 2017). However, crop rotation with non-mycorrhizal crops, such as the Brassica family, reduces the abundance of AMF in the soil and the symbiotic benefits conferred to the crops through the production of antimicrobial isothiocyanates (Valetti et al., 2016). Moreover, some plants are more mycorrhizal-dependent than others. For instance, the maize crop is more mycotrope than soybean (Wang et al., 2016), while wheat is considered a non-mycorrhizal or mycorrhizal plant species depending on the cultivar (Stefani et al., 2020).

Wheat cultivars respond differently to AMF inoculation in terms of growth, root colonization, and carbon for nutrient exchange (García de León et al., 2020). This mechanism is known as a mycorrhizal dependency, whereby AMF presence in the soil affects the growth response of plant species differentially (Kandhasamy et al., 2020). In addition, intercropping various plant species in similar conditions impact the composition of the community and diversity of AMF and the plant diversity. Indeed, AMF is involved in transporting plant assimilates from the dominant species to plant species subordinated through the AMF mycelium network (Egerton-Warburton et al., 2007). Therefore, the presence of AMF impacts the type of vegetation, the relative abundance of plant species, and their diversity (Yadav et al., 2020).

This mechanism operates through a change in the soil microorganisms and soil properties (Liu et al., 2020). Microbial populations in the rhizosphere are known to either interfere with or benefit mycorrhizal establishment (Germida & Walley, 1996). Soil microorganisms can produce compounds that increase root cell permeability and can increase the rates of root exudation. This, in turn, would stimulate mycorrhizal fungal mycelia in the rhizosphere or facilitate root penetration by the fungus. Mainly, mycorrhiza helper bacteria are known to stimulate the mycelial growth of mycorrhizal fungi and/or enhance mycorrhizal formation (Gryndler et al., 2000).

6.2. Impact of Soil Management Practice on AMF

Soil management practices have a significant impact on soil properties and microbial diversity. Mineral fertilizer, chemical pesticides, and herbicides applications are essential for crop nutrition, and they replenish the soil nutrients pool removed or harvested by crops, weed control, and pest management (Rana et al., 2019). However, high- and long-term inorganic fertilizer application reduces the plant's dependency on AMF, subsequently, mycorrhizal diversity, and abundance (Kour et al., 2020). This phenomenon can be explained by the fact that the symbiotic relationship between the plant and AMF is energetically costly for the plant; therefore, when the soil is rich in nutrients, plants allocate fewer carbohydrates to AMF, which ultimately affects the spore development and hyphae production (Tian et al., 2013).

Physical characteristics, such as soil texture, as well as seasonality and soil moisture, are also important drivers of AMF species distribution (Cuenca & Lovera, 2010). Moreover, the composition and structure of AMF communities are affected by soil texture, moisture and aeration (Lekberg et al., 2007). Accounting for confounding effects by differences in soil texture is thus important in field surveys to study the effects of land use on soil fertility at different locations, such as on different farms (Ciccolini et al., 2016a). In addition, organic fertilizer application also has both positive and negative impacts on AMF diversity (Liu et al., 2020). Reduced carbon levels and mineral fertilization (e.g., phosphorus), decline in plant diversity, and certain soil management practices can lead to a reduction in AMF diversity (de Souza et al., 2013).

According to Zhu et al. (2016), organic matter improves the AMF community composition in the rhizosphere of maize. In the same vein, inorganic pesticide and herbicide applications also have both positive and negative effects on AMF. These effects are mediated by the secretion of active substances up taken by plants via root or hyphal from the rhizosphere (Hage-Ahmed et al., 2019). Preceding crops and residual effects of agronomic interventions cause legacies in chemical and biological soil properties (Hallama et al., 2019), including microbial biomass, AMF abundance and diversity (Higo et al., 2019; Lehman et al., 2019) with cascading effects on crop productivity (Manoharan et al., 2017). Soil phosphorus plays the most significant role in regulating plant mycorrhizal symbiosis (Kowalska et al., 2015). Thus, high P application negatively affects root colonization and AMF diversity (Cheng et al., 2013).

On the other side, low fertilizer application optimizes the plant mycorrhizal symbiosis (Rana et al., 2020). For instance, Liu et al. (2015) found that P application in nutrient-deprived soil improves the mycorrhizal-mediated benefits to the plant. In addition to the chemical application, extensive tillage greatly influences the community composition of AMF by reducing mycelium extension, colonization rate, and diversity structure (Zhao et al., 2015). This is due to the effects of tillage on the soil's permeability, texture, and microbial food substrates, which ultimately affect the soil microbiota activity and their habitat type (Wang et al., 2020).

Several studies demonstrated that conservation or zero tillage improves AMF diversity and abundance, resulting in better plant growth (Gu et al., 2020). However, a study carried out in a Mediterranean agro ecosystem found that AMF spore density or extraradical mycelium density is not affected by conventional tillage practices (Curaqueo et al., 2011). These contradicting results indicate that the effects of different chemical applications and tillage practices on AMF still need further studies to understand better the effect of different soil management practices on soil AMF. For instance, azoxystrobin and glyphosate, respectively, fungicide and herbicide, inhibit the spore germination of some AMF species (Buysens et al., 2015).

7. AMF in Sustainable Agriculture

Arbuscular mycorrhizal fungi can improve nutrient acquisition, mainly phosphorus and thereby helps plants to thrive under limited nutrient conditions (Khanday et al., 2016). It is also documented that symbiotic association between plant and mycorrhiza increases the uptake of other nutrients like nitrogen, zinc and potassium and water use of crops as well as enhances symbiotic N-fixation ability (Jansa et al., 2019). Moreover, it is known to modify several aspects of plant physiology including mineral nutrient composition, hormonal balance, C allocation patterns, etc.

The quality of soil depends not only on its physical or chemical properties but also on the diversity and activity of soil biota (Doran and Linn, 1994) that support plants in nutrient and water uptake and contributes to soil ecosystem services. AMF does not only have an impact on plant growth and production but it has been also reported that they improve some soil characteristics, such as soil aggregation, improving soil structure and thereby preventing soil erosion and soil acidity correction (Parihar et al., 2020). The improvement of soil aggregation contributes to the maintenance of good water infiltration rates, good tilth and adequate aeration for plant growth, thus improving soil quality (Wright & Upadhyaya, 1998). Arbuscular mycorrhizal fungi (AMF) and glomalin related soil protein (GRSP) have been suggested to be used as biological indicators of soil quality (Fokom et al., 2012). Consequently, the mycorrhizal symbiotic status changes the chemical composition of root exudates while the development of the fungal soil mycelium serves as a carbon source to rhizosphere microbial communities and introduces physical modifications into the environment surrounding the roots.

The inoculation of mycorrhizal fungi has been shown to improve nodulation and N_2 fixation. The well-known activities of nitrogen-fixing bacteria and phosphate-solubilizing microorganisms improving the bioavailability of the major plant nutrients N and P, are very much enhanced in the rhizosphere of mycorrhizal plants where synergistic interactions of such microorganisms with mycorrhizal fungi have been demonstrated. Management of such interactions is a promising approach for either low-input agricultural technologies (Jeffries & Barea, 2001), or the re-establishment of the natural vegetation in a degraded area (Barea & Jeffries, 1995). AMF has essential participation in the recovery process of degraded areas that are important in the phosphorus, nitrogen and carbon cycle (Veresoglou et al., 2012) and can be influenced by organic amendment (Aleklett & Wallander, 2012). AMF also modify physical soil attributes, forming more stable macroaggregates and increasing resistance to erosion (Rillig, 2004).

Mycorrhiza fungi offer substantial benefits to associated plants like an increase in mineral nutrient uptake, protection against soil-borne pathogens, increased resistance against abiotic stresses like drought and the presence of toxic matters present in soil (Siddiqui and Mahmood, 1995). Where AM fungi contribute to P uptake, the benefit can be measured in terms of savings in fertilizer (Schweiger et al., 2007). There has been little consideration of potential savings in nitrogen fertilizer (Johnson, 2010).

Conclusion

Due to the excessive use of inorganic fertilizers and chemicals, Arbuscular mycorrhizal (AM) fungal symbiosis is the primary element of the agricultural system to improve soil health, in addition to enhancing plant nutrition, growth and health.

AM fungi are found to improve soil physicochemical and biological properties of soil by promoting soil aggregation, soil carbon and nutrients availability, water retention, microbial activities, nitrogen, carbon, and phosphorus cycling, thereby preventing soil erosion, carbon, and nitrogen loss as well as remediation of heavy metals. Moreover, AM fungi provide sustainable soil health by improving defense mechanisms against plant root pathogens, providing availability of essential plant nutrients, biocontrol of pests and play an efficient role in phytoremediation.

Besides, AM fungi also offer sustainable benefits to associated plants such as nutrient availability, exclusively PO_4^{2-} and NO_3^-, improving defense mechanisms against phytopathogens and herbivorous insect pests as well as abiotic stresses drought, high temperature, heavy metal toxicity, and so on. However, the effects of AMF on soil health and fertility are governed by the richness of AMF fungal species, tillage, and soil condition. Hence, efficient utilization of AM fungi improves plant nutrition and soil health for the betterment of sustainable crop yield and acts as a component of an integrated system of sustainable agriculture.

References

Abbott LK, Johnson NC (2017) Mycorrhizal Mediation of Soil: Fertility, Structure and Carbon Storage. Elsevier, Amsterdam, pp 93–105.

Albrechtova J, Latr A, Nedorost L, Pokluda R, Posta K., Vosatka M (2012) Dual inoculation with mycorrhizal and saprotrophic fungi applicable in sustainable cultivation improves the yield and nutritive value of onion. The Scientific World Journal, 2012:374091.

Aleklett K, Wallander H (2012) Effects of organic amendments with various nitrogen levels on arbuscular mycorrhizal fungal growth. Applied Soil Ecology, 60:71-76.

Alimi A, Adeleke R, Moteetee A (2021) Soil environmental factors shape the rhizosphere arbuscular mycorrhizal fungal communities in South African indigenous legumes (Fabaceae). Biodiversitas Journal of Biological Diversity, 22(5):2466-2476.

Allen MF, Sexton JC, Jr Moore TS, Christensen M (1981) Influence of phosphate source on vesicular-arbuscular mycorrhizae of Bouteloua gracilis. New Phytologist, 87(4):687-694.

Augé RM (2001) Water relations, drought and vesicular-arbuscular mycorrhizal symbiosis. Mycorrhiza, 11(1):3-42.

Bai C, He X, Tang H, Shan B, Zhao L (2009) Spatial distribution of arbuscular mycorrhizal fungi, glomalin and soil enzymes under the canopy of Astragalus adsurgens Pall. in the Mu Us sandland, China. Soil Biology and Biochemistry, 41(5):941-947.

Bakhshandeh S, Corneo PE, Mariotte P, Kertesz MA, Dijkstra FA (2017) Effect of crop rotation on mycorrhizal colonization and wheat yield under different fertilizer treatments. Agriculture, Ecosystems & Environment, 247:130-136.

Baltrus DA (2017) Adaptation, specialization, and coevolution within phytobiomes. Current Opinion in Plant Biology, 38:109-116.

Barea JM (2000) Rhizosphere and mycorrhiza of field crops. In: Toutant JP, Balazs E, Galante E, Lynch JM, Schepers JS, Werner D & Werry PA (Eds.) Biological Resource

Management: Connecting Science and Policy (OECD) (pp 110–125) INRA, Editions and Springer.
Barea JM, Azco'n-Aguilar C, Azcon R (1997) Interactions between mycorrhizal fungi and rhizosphere microorganism within the context of sustainable soil-plant systems. In: Gange AC, Brown VK (eds.) Multitrophic Interactions in Terrestrial Systems. Cambridge University Press, Cambridge, pp 65–77.
Barea JM, Azcón R, Azcón-Aguilar C (1992) Vesicular-arbuscular mycorrhizal fungi in nitrogen-fixing systems. In: Norris JR, Read DJ & Varma AK (Eds.) Methods in Microbiology (pp 391–416). Academic Press, London.
Barea JM, Azcón R, Azcón-Aguilar C (2002) Mycorrhizosphere interactions to improve plant fitness and soil quality. Antonie van Leeuwenhoek, 81(1):343-351.
Barea JM, Jeffries P (1995) Arbuscular mycorrhizas in sustainable soil plant systems. In: Hock B & Varma A (Eds.) Mycorrhiza Structure Function, Molecular Biology and Biotechnology (pp 521–559). Springer-Verlag, Heidelberg.
Bearden BN, Petersen L (2000) Influence of arbuscular mycorrhizal fungi on soil structure and aggregate stability of a vertisol. Plant and Soil, 218(1):173-183.
Bedini S, Pellegrino E, Avio L, Pellegrini S, Bazzoffi P, Argese E, Giovannetti M (2009) Changes in soil aggregation and glomalin-related soil protein content as affected by the arbuscular mycorrhizal fungal species Glomus mosseae and Glomus intraradices. Soil Biology and Biochemistry, 41(7):1491-1496.
Behera B, Singdevsachan SK, Mishra R, Dutta S, Thatoi H (2014) Diversity, mechanism and biotechnology of phosphate solubilising microorganism in mangrove—a review. Biocatalysis and Agricultural Biotechnology, 3(2):97-110.
Bronick CJ, Lal R (2005) Soil structure and management: a review. Geoderma, 124(1-2):3-22.
Buysens C, Dupré de Boulois H, Declerck S (2015) Do fungicides used to control Rhizoctonia solani impact the non-target arbuscular mycorrhizal fungus Rhizophagus irregularis? Mycorrhiza, 25(4): 277-288.
Cheng Y, Ishimoto K, Kuriyama Y, Osaki M, Ezawa T (2013) Ninety-year-, but not single, application of phosphorus fertilizer has a major impact on arbuscular mycorrhizal fungal communities. Plant and Soil, 365(1):397-407.
Ciccolini V, Bonari E, Ercoli L, Pellegrino E (2016) Phylogenetic and multivariate analyses to determine the effect of agricultural land-use intensification and soil physico-chemical properties on N-cycling microbial communities in drained Mediterranean peaty soils. Biology and Fertility of Soils, 52(6):811-824.
Cuenca G, Lovera M (2010) Seasonal variation and distribution at different soil depths of arbuscular mycorrhizal fungi spores in a tropical sclerophyllous shrubland. Botany, 88(1):54-64.
Curaqueo G, Barea JM, Acevedo E, Rubio R, Cornejo P, Borie F (2011) Effects of different tillage system on arbuscular mycorrhizal fungal propagules and physical properties in a Mediterranean agroecosystem in central Chile. Soil and Tillage Research, 113(1):11-18.
de Novais CB, Avio L, Giovannetti M, de Faria SM, Siqueira JO, Sbrana C (2019) Interconnectedness, length and viability of arbuscular mycorrhizal mycelium as affected by selected herbicides and fungicides. Applied Soil Ecology, 143:144-152.

De Souza R, Da Silva D, De Mello C, Goto B, Da Silva F, Sampaio E, Maia L (2013) Arbuscular mycorrhizal fungi in revegetated mined dunes. Land Degradation & Development, 24(2):147-155.

Demenois J, Rey F, Ibanez T, Stokes A, Carriconde F (2018) Linkages between root traits, soil fungi and aggregate stability in tropical plant communities along a successional vegetation gradient. Plant and Soil, 424(1):319-334.

Dewitte O, Jones A, Spaargaren O, Breuning-Madsen H, Brossard M, Dampha A. Jones R (2013) Harmonisation of the soil map of Africa at the continental scale. Geoderma, 211:138-153.

Dobbelaere S, Vanderleyden J, Okon Y (2003) Plant growth-promoting effects of diazotrophs in the rhizosphere. Critical Reviews in Plant Sciences, 22(2):107-149.

Doran JW, Parkin TB (1994) Defining and assessing soil quality. Defining Soil Quality for a Sustainable Environment, 35:1-21.

Drigo B, Pijl AS, Duyts H, Kielak AM, Gamper HA, Houtekamer MJ, Veen JA (2010) Shifting carbon flow from roots into associated microbial communities in response to elevated atmospheric CO_2. Proceedings of the National Academy of Sciences, 107(24):10938-10942.

Egerton-Warburton LM, Querejeta JI, Allen MF (2007) Common mycorrhizal networks provide a potential pathway for the transfer of hydraulically lifted water between plants. Journal of Experimental Botany, 58(6):1473-1483.

Endlweber K, Scheu S (2007) Interactions between mycorrhizal fungi and Collembola: effects on root structure of competing plant species. Biology and Fertility of Soils, 43(6):741-749.

Etesami H, Jeong BR (2021) Contribution of arbuscular mycorrhizal fungi, phosphate–solubilizing bacteria, and silicon to P uptake by plant: a review. Frontiers in Plant Science, 12:1355.

Evelin H, Kapoor R, Giri B (2009) Arbuscular mycorrhizal fungi in alleviation of salt stress: a review. Annals of Botany 104(7):1263-1280.

Finlay RD (2008) Ecological aspects of mycorrhizal symbiosis: with special emphasis on the functional diversity of interactions involving the extraradical mycelium. Journal of Experimental Botany, 59(5):1115-1126.

Fokom R, Adamou S, Teugwa M, Boyogueno AB, Nana W, Ngonkeu M, Zollo PA (2012) Glomalin related soil protein, carbon, nitrogen and soil aggregate stability as affected by land use variation in the humid forest zone of South Cameroon. Soil and Tillage Research, 120:69-75.

Frank AB (1885) Uber di auf werzelsymbiose beruhende Ernahrung gewisser Baume durch unterirdischeplize [On the nutrition of certain trees based on root symbiosis by subterranean mushrooms]. Ber Dtsch Bot Ges 3:128–145.

Garbisu C, Alkorta I (2001) Phytoextraction: a cost-effective plant-based technology for the removal of metals from the environment. Bioresource Technology, 77(3):229-236.

Garcia de Leon D, Vahter T, Zobel M, Koppel M, Edesi L, Davison J, Oja J (2020) Different wheat cultivars exhibit variable responses to inoculation with arbuscular mycorrhizal fungi from organic and conventional farms. PloS one, 15(5):e0233878.

Garcia K, Doidy J, Zimmermann SD, Wipf D, Courty PE (2016) Take a trip through the plant and fungal transportome of mycorrhiza. Trends in Plant Science, 21(11):937-950.

Gaur A, Adholeya A (2004) Prospects of arbuscular mycorrhizal fungi in phytoremediation of heavy metal contaminated soils. Current Science, 86:528-534.

Germida J, Walley F (1996) Plant growth-promoting rhizobacteria alter rooting patterns and arbuscular mycorrhizal fungi colonization of field-grown spring wheat. Biology and Fertility of Soils, 23(2):113-120.

Gessner MO, Swan CM, Dang CK, McKie BG, Bardgett RD, Wall DH, Hättenschwiler S (2010) Diversity meets decomposition. Trends in Ecology & Evolution, 25(6): 372-380.

Gianinazzi S, Gollotte A, Binet MN, van Tuinen D, Redecker D, Wipf D (2010) Agroecology: the key role of arbuscular mycorrhizas in ecosystem services. Mycorrhiza, 20(8):519-530.

Göhre V, Paszkowski U (2006) Contribution of the arbuscular mycorrhizal symbiosis to heavy metal phytoremediation. Planta, 223(6):1115-1122.

Gryndler M (2000) Interactions of arbuscular mycorrhizal fungi with other soil organisms. In Arbuscular Mycorrhizas: Physiology and Function (pp. 239-262): Springer.

Gu S, Wu S, Guan Y, Zhai C, Zhang Z, Bello A, Yang W (2020) Arbuscular mycorrhizal fungal community was affected by tillage practices rather than residue management in black soil of northeast China. Soil and Tillage Research, 198:104552.

Hage-Ahmed K, Rosner K, Steinkellner S (2019) Arbuscular mycorrhizal fungi and their response to pesticides. Pest Management Science, 75(3):583-590.

Hallama M, Pekrun C, Lambers H, Kandeler E (2019) Hidden miners–the roles of cover crops and soil microorganisms in phosphorus cycling through agroecosystems. Plant and Soil, 434(1):7-45.

Hamel C, Plenchette C (2017) Implications of past, current, and future agricultural practices for mycorrhiza-mediated nutrient flux. In Mycorrhizal Mediation of Soil, eds N. C. Johnson, C. Gehring, and J. Jansa (Amsterdam: Elsevier), 175–186.

Haselwandter K (2008) Structure and function of siderophores produced by mycorrhizal fungi. Mineralogical Magazine, 72(1):61-64.

Haselwandter K, Häninger G, Ganzera M, Haas H, Nicholson G, Winkelmann G (2013) Linear fusigen as the major hydroxamate siderophore of the ectomycorrhizal Basidiomycota Laccaria laccata and Laccaria bicolor. Biometals, 26(6):969-979.

Hashem A, Alqarawi AA, Radhakrishnan R, Al-Arjani ABF, Aldehaish HA, Egamberdieva D, Abd_Allah EF (2018) Arbuscular mycorrhizal fungi regulate the oxidative system, hormones and ionic equilibrium to trigger salt stress tolerance in Cucumis sativus L. Saudi Journal of Biological Sciences, 25(6):1102-1114.

Higo M, Tatewaki Y, Gunji K, Kaseda A, Isobe K (2019) Cover cropping can be a stronger determinant than host crop identity for arbuscular mycorrhizal fungal communities colonizing maize and soybean. Peer J, 7:e6403.

Hu D, Baskin JM, Baskin CC, Wang Z, Zhang S, Yang X, Huang Z (2019) Arbuscular mycorrhizal symbiosis and achene mucilage have independent functions in seedling growth of a desert shrub. Journal of Plant Physiology, 232:1-11.

Jansa J, Forczek ST, Rozmoš M, Püschel D, Bukovská P, Hršelová H (2019) Arbuscular mycorrhiza and soil organic nitrogen: network of players and interactions. Chemical and Biological Technologies in Agriculture, 6(1):1-10.

Jeffries P Barea JM (2001) Arbuscular Mycorrhiza – a key component of sustainable plant-soil ecosystems. In: Hock B (Ed.) The Mycota. Vol. IX Fungal Associations (pp 95–113). Springer-Verlag, Berlin, Heidelberg.

Johnson NC (2010) Resource stoichiometry elucidates the structure and function of arbuscular mycorrhizas across scales. New Phytologist, 185(3):631-647.

Jurinak J, Dudley L, Allen M, Knight W (1986) The role of calcium oxalate in the availability of phosphorus in soils of semiarid regions: a thermodynamic study. Soil Science, 142(5):255-261.

Kalayu G (2019) Phosphate solubilizing microorganisms: promising approach as biofertilizers. International Journal of Agronomy, 2019.

Kandhasamy N, Ravichandran KR, Thangavelu M (2020) Interactive influence of soil and plant genotypes on mycorrhizal dependency in finger millet. Journal of Soil Science and Plant Nutrition, 20(3):1287-1297.

Khanday M, Bhat RA, Haq S, Dervash MA, Bhatti AA, Nissa M, Mir MR (2016) Arbuscular mycorrhizal fungi boon for plant nutrition and soil health. In: Hakeem, KR, Akhtar, J, & Sabir, M. (eds.) Soil Science: Agricultural and Environmental Prospective. Springer, Cham, pp 317–332.

Kour D, Rana KL, Kaur T, Yadav N, Halder SK, Yadav AN (2020) Potassium solubilizing and mobilizing microbes: biodiversity, mechanisms of solubilization, and biotechnological implication for alleviations of abiotic stress. In: New and Future Developments in Microbial Biotechnology and Bioengineering, (eds.) A. A. Rastegari, A. N. Yadav, and N. Yadav (Amsterdam: Elsevier), 177–202.

Kowalska I, Konieczny A, Gąstoł M, Sady W, Hanus-Fajerska E (2015) Effect of mycorrhiza and phosphorus content in nutrient solution on the yield and nutritional status of tomato plants grown on rockwool or coconut coir. Agricultural and Food Science, 24(1):39-51.

Krishna K, Bagyaraj D (1984) Growth and nutrient uptake of peanut inoculated with the mycorrhizal fungus Glomus fasciculatum compared with non-inoculated ones. Plant and Soil, 77(2):405-408.

Lambers H, Raven JA, Shaver GR, Smith SE (2008) Plant nutrient-acquisition strategies change with soil age. Trends in Ecology & Evolution, 23(2):95-103.

Lapeyrie F (1988) Oxalate synthesis from soil bicarbonate by the mycorrhizal fungusPaxillus involutus. Plant and Soil, 110(1):3-8.

Larimer AL, Clay K, Bever JD (2014) Synergism and context dependency of interactions between arbuscular mycorrhizal fungi and rhizobia with a prairie legume. Ecology, 95(4):1045-1054.

Leake J, Johnson D, Donnelly D, Muckle G, Boddy L, Read D (2004) Networks of power and influence: the role of mycorrhizal mycelium in controlling plant communities and agroecosystem functioning. Canadian Journal of Botany, 82(8): 1016-1045.

Lehman RM, Osborne SL, Taheri WI, Buyer JS, Chim BK (2019) Comparative measurements of arbuscular mycorrhizal fungal responses to agricultural management practices. Mycorrhiza, 29(3):227-235.

Lehmann J, Hansel CM, Kaiser C, Kleber M, Maher K, Manzoni S, Torn MS (2020) Persistence of soil organic carbon caused by functional complexity. Nature Geoscience, 13(8):529-534.

Lekberg Y, Koide RT, Rohr JR, Aldrich-wolfe L, Morton JB (2007) Role of niche restrictions and dispersal in the composition of arbuscular mycorrhizal fungal communities. Journal of Ecology, 95(1):95-105.

Liu A, Hamel C, Hamilton R, Ma B, Smith D (2000) Acquisition of Cu, Zn, Mn and Fe by mycorrhizal maize (Zea mays L.) grown in soil at different P and micronutrient levels. Mycorrhiza, 9(6):331-336.

Liu J, Zhang J, Li D, Xu C, Xiang X (2020) Differential responses of arbuscular mycorrhizal fungal communities to mineral and organic fertilization. Microbiology Open, 9(1):e00920.

Liu L, Zhang T, Gilliam FS, Gundersen P, Zhang W, Chen H, Mo J (2013) Interactive effects of nitrogen and phosphorus on soil microbial communities in a tropical forest. PloSOne, 8(4):e61188.

Liu Y, Johnson NC, Mao L, Shi G, Jiang S, Ma X, Feng H (2015) Phylogenetic structure of arbuscular mycorrhizal community shifts in response to increasing soil fertility. Soil Biology and Biochemistry, 89:196-205.

Lucy M, Reed E, Glick BR (2004) Applications of free living plant growth-promoting rhizobacteria. Antonie van Leeuwenhoek, 86(1):1-25.

Mahanta D, Rai R, Dhar S, Varghese E, Raja A, Purakayastha T (2018) Modification of root properties with phosphate solubilizing bacteria and arbuscular mycorrhiza to reduce rock phosphate application in soybean-wheat cropping system. Ecological Engineering, 111:31-43.

Manoharan L, Rosenstock NP, Williams A, Hedlund K (2017) Agricultural management practices influence AMF diversity and community composition with cascading effects on plant productivity. Applied Soil Ecology, 115:53-59.

Manoharan P, Shanmugaiah V, Balasubramanian N, Gomathinayagam S, Sharma MP, Muthuchelian K (2010) Influence of AM fungi on the growth and physiological status of Erythrina variegata Linn. grown under different water stress conditions. European Journal of Soil Biology, 46(2):151-156.

Mcgonigle TP, Miller MH (1996) Development of fungi below ground in association with plants growing in disturbed and undisturbed soils. Soil Biology and Biochemistry, 28(3):263-269.

Miller R, Jastrow J (1990) Hierarchy of root and mycorrhizal fungal interactions with soil aggregation. Soil Biology and Biochemistry, 22(5):579-584.

Miller RM, Jastrow JD (1992) The role of mycorrhizal fungi in soil conservation. Mycorrhizae in Sustainable Agriculture, 54:29-44.

Miller RM, Jastrow JD (2000) Mycorrhizal fungi influence soil structure. In: Kapulnik Y & Douds DD Jr (Eds.) Arbuscular Mycorrhizas: Physiology and Function (pp 3–18). Kluwer Academic Publishers, Dordrecht, The Netherlands.

Miransari M (2010) Contribution of arbuscular mycorrhizal symbiosis to plant growth under different types of soil stress. Plant Biology, 12(4):563-569.

Mubekaphi C (2019) Soil organic carbon, glomalin related soil protein and related physical properties after 15 years of different management practices in a subtropical region of

South Africa (Doctoral thesis). University of KwaZulu- Natal, Pietermaritzburg, South Africa.

Nichols K, Wright S (2004) Contributions of soil fungi to organic matter in agricultural soils. Functions and Management of Soil Organic Matter in Agroecosystems, 179-198.

Nicolson T (1967) Vesicular-arbuscular mycorrhiza—a universal plant symbiosis. Science Progress (1933):561-581.

Nielsen UN, Ayres E, Wall DH, Bardgett RD (2011) Soil biodiversity and carbon cycling: a review and synthesis of studies examining diversity–function relationships. European Journal of Soil Science, 62(1):105-116.

Oades JM (1984) Soil organic matter and structural stability: mechanisms and implications for management. Plant and Soil, 76(1):319-337.

Oldfield EE, Wood SA, Bradford MA (2018) Direct effects of soil organic matter on productivity mirror those observed with organic amendments. Plant and Soil, 423(1):363-373.

Othman R, Panhwar QA (2014) Phosphate-solubilizing bacteria improves nutrient uptake in aerobic rice. In Phosphate Solubilizing Microorganisms (pp. 207-224): Springer.

Parihar M, Rakshit A, Meena VS, Gupta VK, Rana K, Choudhary M, Bisht JK (2020) The potential of arbuscular mycorrhizal fungi in C cycling: a review. Archives of Microbiology, 202(7):1581-1596.

Pellegrino E, Gamper HA, Ciccolini V, Ercoli L (2020) Forage rotations conserve diversity of arbuscular mycorrhizal fungi and soil fertility. Frontiers in Microbiology, 2969.

Powell JR, Rillig MC (2018) Biodiversity of arbuscular mycorrhizal fungi and ecosystem function. New Phytologist, 220(4):1059-1075.

Rana KL, Kour D, Kaur T, Sheikh I, Yadav AN, Kumar V, Dhaliwal HS (2020) Endophytic microbes from diverse wheat genotypes and their potential biotechnological applications in plant growth promotion and nutrient uptake. Proceedings of the National Academy of Sciences, India Section B: Biological Sciences, 90(5):969-979.

Rana KL, Kour D, Sheikh I, Dhiman A, Yadav N, Yadav AN (2019) Endophytic fungi: biodiversity, ecological significance, and potential industrial applications," in Recent Advancement in White Biotechnology Through Fungi, eds. A. Yadav, S. Mishra, S. Singh, and A. Gupta (Cham: Springer), 1–62.

Rillig MC (2004) Arbuscular mycorrhizae, glomalin, and soil aggregation. Canadian Journal of Soil Science, 84(4):355-363.

Rillig MC, Caldwell BA, Wösten HA, Sollins P (2007) Role of proteins in soil carbon and nitrogen storage: controls on persistence. Biogeochemistry, 85(1):25-44.

Rillig MC, Wagner M, Salem M, Antunes PM, George C, Ramke HG, Antonietti M (2010) Material derived from hydrothermal carbonization: effects on plant growth and arbuscular mycorrhiza. Applied Soil Ecology, 45(3):238-242.

Saba H, Jyoti P, Neha S (2013) Mycorrhizae and phytochelators as remedy in heavy metal contaminated land remediation. Int Res J Environ Sci, 2(1):74-78.

Sanderman J, Hengl T, Fiske GJ (2017) Soil carbon debt of 12,000 years of human land use. Proceedings of the National Academy of Sciences, 114(36):9575-9580.

Schloter M, Nannipieri P, Sørensen SJ, van Elsas JD (2018) Microbial indicators for soil quality. Biology and Fertility of Soils, 54(1):1-10.

Schoeneberger M, Volk R, Davey C (1989) Selection of reference plants and methods for estimating N_2 fixation in mycorrhizal leguminous plants. Soil Science Society of America Journal, 53(5):1425-1429.

Schweiger PF, Robson AD, Barrow NJ, Abbott LK (2007) Arbuscular mycorrhizal fungi from three genera induce two-phase plant growth responses on a high P-fixing soil. Plant and Soil, 292(1):181-192.

Shen J, Yuan L, Zhang J, Li H, Bai Z, Chen X, Zhang F (2011) Phosphorus dynamics: from soil to plant. Plant Physiology, 156(3):997-1005.

Siddiqui ZA, Mahmood I (1995) Role of plant symbionts in nematode management: a review. Bioresource Technology, 54(3):217-226.

Singh AK, Zhu X, Chen C, Wu J, Yang B, Zakari S, Liu W (2020) The role of glomalin in mitigation of multiple soil degradation problems. Critical Reviews in Environmental Science and Technology, 52(9):1604-1638.

Six J, Carpentier A, van Kessel C, Merckx R, Harris D, Horwath WR, Lüscher A (2001) Impact of elevated CO_2 on soil organic matter dynamics as related to changes in aggregate turnover and residue quality. Plant and Soil, 234(1):27-36.

Skujins J, Allen MF (1986) Use of mycorrhizae for land rehabilitation. MIRCEN J 2:161–176.

Smith SE, Read DJ (2008) Microbial Symbiosis, 3rd edn. Academic, London.

Smith SE, Smith FA (2012) Fresh perspectives on the roles of arbuscular mycorrhizal fungi in plant nutrition and growth. Mycologia, 104(1):1-13.

Stefani F, Dupont S, Laterrière M, Knox R, Ruan Y, Hamel C, Hijri M (2020) Similar arbuscular mycorrhizal fungal communities in 31 durum wheat cultivars (Triticum turgidum L. var. durum) under field conditions in Eastern Canada. Frontiers in Plant Science, 11:1206.

Taktek S, St-Arnaud M, Piché Y, Fortin JA, Antoun H (2017) Igneous phosphate rock solubilization by biofilm-forming mycorrhizobacteria and hyphobacteria associated with Rhizoglomus irregulare DAOM 197198. Mycorrhiza, 27(1):13-22.

Tarafdar J, Marschner H (1994) Phosphatase activity in the rhizosphere and hyphosphere of VA mycorrhizal wheat supplied with inorganic and organic phosphorus. Soil Biology and Biochemistry, 26(3):387-395.

Taylor J, Harrier LA (2001) A comparison of development and mineral nutrition of micropropagated Fragaria× ananassa cv. Elvira (strawberry) when colonised by nine species of arbuscular mycorrhizal fungi. Applied Soil Ecology, 18(3):205-215.

Tian H, Drijber R, Zhang J, Li X (2013) Impact of long-term nitrogen fertilization and rotation with soybean on the diversity and phosphorus metabolism of indigenous arbuscular mycorrhizal fungi within the roots of maize (Zea mays L.). Agriculture, Ecosystems & Environment, 164:53-61.

Tisdall J, Smith S, Rengasamy P (1997) Aggregation of soil by fungal hyphae. Soil Research, 35(1):55-60.

Valetti L, Iriarte L, Fabra A (2016) Effect of previous cropping of rapeseed (Brassica napus L.) on soybean (Glycine max) root mycorrhization, nodulation, and plant growth. European Journal of Soil Biology, 76:103-106.

Van Aarle IM, Rouhier H, Saito M (2002) Phosphatase activities of arbuscular mycorrhizal intraradical and extraradical mycelium, and their relation to phosphorus availability. Mycological Research, 106(10):1224-1229.

Veresoglou, SD, Chen, B, & Rillig, MC. (2012). Arbuscular mycorrhiza and soil nitrogen cycling. Soil Biology and Biochemistry, 46, 53-62.

Volpin H, Elkind Y, Okon Y, Kapulnik Y (1994) A vesicular arbuscular mycorrhizal fungus (Glomus intraradies) induces a defense response in alfalfa roots. Plant Physiology, 104(2):683-689.

Wang G, Sheng L, Zhao D, Sheng J, Wang X, Liao H (2016) Allocation of nitrogen and carbon is regulated by nodulation and mycorrhizal networks in soybean/maize intercropping system. Frontiers in Plant Science, 7:1901.

Wang X, Pan Q, Chen F, Yan X, Liao H (2011) Effects of co-inoculation with arbuscular mycorrhizal fungi and rhizobia on soybean growth as related to root architecture and availability of N and P. Mycorrhiza, 21(3):173-181.

Wang Z, Li Y, Li T, Zhao D, Liao Y (2020) Conservation tillage decreases selection pressure on community assembly in the rhizosphere of arbuscular mycorrhizal fungi. Science of the Total Environment, 710:136326.

Wei Y, Zhao Y, Fan Y, Lu Q, Li M, Wei Q, Wei Z (2017) Impact of phosphate-solubilizing bacteria inoculation methods on phosphorus transformation and long-term utilization in composting. Bioresource Technology, 241:134-141.

Willis A, Rodrigues B, Harris PJ (2013) The ecology of arbuscular mycorrhizal fungi. Critical Reviews in Plant Sciences, 32(1):1-20.

Wright SF, Upadhyaya A (1998) A survey of soils for aggregate stability and glomalin, a glycoprotein produced by hyphae of arbuscular mycorrhizal fungi. Plant and Soil, 198(1):97-107.

Xu J, Liu S, Song S, Guo H, Tang J, Yong JW, Chen X (2018) Arbuscular mycorrhizal fungi influence decomposition and the associated soil microbial community under different soil phosphorus availability. Soil Biology and Biochemistry, 120:181-190.

Xu M, Li X, Cai X, Li X, Christie P, Zhang J (2017) Land use alters arbuscular mycorrhizal fungal communities and their potential role in carbon sequestration on the Tibetan Plateau. Scientific Reports, 7(1):1-11.

Yadav R, Ror P, Rathore P, Ramakrishna W (2020) Bacteria from native soil in combination with arbuscular mycorrhizal fungi augment wheat yield and biofortification. Plant Physiology and Biochemistry, 150:222-233.

Yang Y, He C, Huang L, Ban Y, Tang M (2017) The effects of arbuscular mycorrhizal fungi on glomalin-related soil protein distribution, aggregate stability and their relationships with soil properties at different soil depths in lead-zinc contaminated area. PloS one, 12(8):e0182264.

Zhang L, Shi N, Fan J, Wang F, George TS, Feng G (2018) Arbuscular mycorrhizal fungi stimulate organic phosphate mobilization associated with changing bacterial community structure under field conditions. Environmental Microbiology, 20(7):2639-2651.

Zhang L, Xu M, Liu Y, Zhang F, Hodge A, Feng G (2016) Carbon and phosphorus exchange may enable cooperation between an arbuscular mycorrhizal fungus and a phosphate-solubilizing bacterium. New Phytologist, 210(3):1022-1032.

Zhao R, Guo W, Bi N, Guo J, Wang L, Zhao J, Zhang J (2015) Arbuscular mycorrhizal fungi affect the growth, nutrient uptake and water status of maize (Zea mays L.) grown in two types of coal mine spoils under drought stress. Applied Soil Ecology, 88:41-49.

Zhu C, Ling N, Guo J, Wang M, Guo S, Shen Q (2016) Impacts of fertilization regimes on arbuscular mycorrhizal fungal (AMF) community composition were correlated with organic matter composition in maize rhizosphere soil. Frontiers in Microbiology, 7:1840.

Zhu J, Li M, Whelan M (2018) Phosphorus activators contribute to legacy phosphorus availability in agricultural soils: A review. Science of the Total Environment, 612:522-537.

Zhu YG, Miller RM (2003) Carbon cycling by arbuscular mycorrhizal fungi in soil–plant systems. Trends in Plant Science, 8(9):407-409.

Index

D

N

O

P

R

S

About the Editors

Prof. Dr. Qiang-Sheng Wu
College of Horticulture and Gardening,
Yangtze University, Jingzhou, China
E-mail: wuqiangsheng@yangtzeu.edu.cn

Prof. Dr. Qiang-Sheng Wu received his Ph.D. from Huazhong Agricultural University in 2006 and is now a professor at the College of Horticulture and Gardening, Yangtze University, along with the director of the Institute of Root Biology, Yangtze University, as well as the invited professor of Univerzita Hradec Králové. He is working on the research of arbuscular mycorrhizal fungi, focusing on the physiological and molecular mechanisms regarding increased drought tolerance of plants by arbuscular mycorrhizal fungi. Prof. Wu is also clarifying the function of arbuscular mycorrhizal fungi-released glomalin and the field application and evaluation of glomalin and arbuscular mycorrhizal fungi.

To date, he has published more than 150 papers as first and/or corresponding author in popular journals such as *Mycorrhiza*, *Frontiers in Microbiology*, *Fungal Ecology*, *Tree Physiology*, *Applied Soil Ecology*, *Journal of Integrative Plant Biology*, *Soil and Tillage Research*, and so on. He has also edited six books published by Springer or NOVA Science Publishers.

Prof. Wu is also the editor of seven popular journals including *Frontiers in Plant Science*, *Chemical and Biological Technologies in Agriculture*, *Notulae Botanicae Horti Agrobotanici Cluj-Napoca*, *Phyton-International Journal of Experimental Botany*, *Agriculture*, *Horticulturae*, and *BIOCELL*.

Prof. Ying-Ning Zou
College of Horticulture and Gardening,
Yangtze University, Jingzhou, China
E-mail: zouyingning@163.com

Prof. Ying-Ning Zou received her master's degree in Huazhong Agricultural University in 2006. She is now a professor at Yangtze University. She is engaged in the research of arbuscular mycorrhizal fungi in fruit trees, with emphasis on the relationship between soil mycorrhizal fungi and drought tolerance of plants. She has published more than 30 papers in the international popular journals, such as *Plant Biology*, *Frontiers in Plant Science*, *Plant Physiology and Biochemistry*, *Applied Soil Ecology*, etc. She is invited as the guest editor in *Forests*.

Prof. Dr. Yue-Jun He
Forestry College,
Guizhou University, Guiyang, China
E-mail: hyj1358@163.com

Prof. Dr. Yue-Jun He received his Ph.D. from Southwest University in 2012 and is now a professor at the Forestry College, Guizhou University, along with the director of the Institute of Research Center of Forest Ecology, Guizhou University. He is working on the research of arbuscular mycorrhizal fungi, focusing on the ecological process and mechanisms regarding plant individuals' adaptability and community stability regulated by arbuscular mycorrhizal fungi in the degraded karst habitat, such as plant invasion competition, plant-soil feedback, habitat heterogeneity, nutrient transfer among plant individuals, litter decomposition, drought tolerance of plants.

To date, Prof. Dr. Yue-Jun He has published more than 70 papers as the first or the corresponding author in popular journals such as *Frontiers in Ecology and Evolution*, *Frontiers in Plant Science*, *Journal of Plant Ecology*, *Agronomy*, *Forests*, *Environmental Science and Pollution Research*, *Acta Ecologica Sinica*, and so on. He has also edited 3 books published by China Science Press and China Forestry Publishing House.

Prof. Nong Zhou
College of Food and Biological Engineering,
Chongqing Three Gorges University, Chongqing, China
E-mail: erhaizn@126.com

Prof. Nong Zhou received his master's degree from Southwest University in 2008 and is studying for a Ph.D. at the College of Pharmacy, Nanjing University of Chinese Medicine. He is now a professor at the Chongqing Three Gorges University, as well as the deputy director of the Scientific Research Office of Chongqing Three Gorges University. He is engaged in the research of arbuscular mycorrhizal fungi of medicinal plants, focusing on the influence of arbuscular mycorrhizal fungi on the growth of *Paris polyphylla* var. yunnanensis.

To date, he has published more than 190 papers as first and/or corresponding author in popular journals such as *Horticulturae*, *Frontiers in Bioengineering* and *Biotechnology*, and so on. Prof. Zhou has edited ten textbooks or monographs. He is also invited as the editor in *Horticulturae*.